Indians, Oil, and Politics

Indians, Oil, and Politics

A Recent History of Ecuador

Allen Gerlach

A Scholarly Resources Inc. Imprint
Wilmington, Delaware

First published 2003
Printed and bound in the United States of America

Scholarly Resources Inc.
104 Greenhill Avenue
Wilmington, DE 19805-1897
www.scholarly.com

All illustrations are courtesy of the photo archives of Guillermo Granja of R.P.A., Retratos Photo Agency, Quito, Ecuador

Library of Congress Cataloging-in-Publication Data

Gerlach, Allen, 1945–
Indians, oil, and politics : a recent history of Ecuador / Allen Gerlach.
p. cm. — (Latin American silhouettes)
Includes bibliographical references and index.
ISBN 0-8420-5107-4 (alk. paper) — ISBN 0-8420-5108-2 (pbk. : alk. paper)
1. Ecuador—Politics and government—1984– 2. Petroleum industry and trade—Ecuador—Finance. 3. Ecuador—Economic conditions. 4. Political stability—Ecuador—History—20th century. 5. Indians of South America—Ecuador—Government relations. 6. Peasant uprisings—Ecuador. I. Title. II. Series.

F3738.2 .G47 2002
986.607'4—dc21 2002030656

∞ The paper used in this publication meets the minimum requirements of the American National Standard for permanence of paper for printed library materials, Z39.48, 1984.

For my daughter, Marty Elizabeth Gerlach, as she notices history's imprints in her own footsteps and comes to appreciate how those who've gone before condition her present and future, and for my wife, Martha Alice Gerlach, who continues to elevate, strengthen, and sustain my spirit. And to the memory of a fine friend and mentor, Edwin Lieuwen, 1923–1988.

Por todo lo que les debo.

About the Author

Allen Gerlach holds a doctorate in history and a law degree from the University of New Mexico. He lived in Peru and Ecuador for several years, and taught Ecuadorean, Andean, and Latin American history at the Centro Andino in Quito as well as at the University of New Mexico in Albuquerque and New Mexico State University in Las Cruces. He has written primarily on Latin America for newspapers, magazines, and academic journals. Currently an attorney, he lives in Albuquerque with his wife and daughter.

Acknowledgments

Every thought and act owes its complexion to the acts of your dead and living brothers. William James would doubtless have agreed that no book is truly written alone. My intellectual and other debts are quite large to individuals who can never adequately be given the credit or appreciation they deserve. I realize this work could not have been done without the inspiration, instruction, and encouragement of too many to mention. My earliest and best teachers, Reuben and Marie Gerlach, imparted early to me in their own ways that for both better and worse the past molds us all, and all too well. I will always be indebted to my brother, Larry R. Gerlach, whose enthusiasm for history and example of excellence led me into the discipline in the first place.

I owe much to other relatives of another sort, my Latin American history godfathers, each an outstanding scholar and mentor. Michael C. Meyer, now of the University of Arizona and previously of the University of Nebraska, and Edwin Lieuwen and France V. Scholes of the University of New Mexico (now deceased), represented different generations, interests, and approaches but were firmly linked by excellence and goodwill. For decades they honored me with their generous guidance, patience, and friendship.

I want to express my appreciation to the people of Ecuador for their hospitality and warmth since 1970, and for their patience in providing interviews, opinions, and insights into their multi-ethnic society with its deep differences and profoundly complex problems. Thanks to those who gave so much: congressmen and congresswomen, presidents and vice presidents, judges, politicians, soldiers, social movement members and directors, writers, reporters, teachers, students, pollsters, business and labor leaders, clerics, and countless others of diverse outlooks, occupations, and conditions. For what I may have managed to get right, the credit belongs to them; if there are errors, the fault is mine. The kind and considerate officials and staff of CONAIE, the Confederación de Nacionalidades Indígenas del Ecuador, merit special thanks for aiding my efforts to grasp the Indian movement's origins, evolution, and orientation. *Gracias y un abrazo fuerte* to Guillermo Granja, photographer and traveling companion. Finally, I want to commend countless university students in Quito and New Mexico for asking the difficult questions that stimulated intellectual growth and curiosity.

Contents

Style Note, **xi**

Introduction, **xiii**

Contrasting Voices, **xvii**

Chapter 1 The Land and the People, **1**

Chapter 2 Historical Background to 1972, **15**

Chapter 3 The Oil Era, **33**

Chapter 4 The Emergence of the Indian Movement, **51**

Chapter 5 Bucaram, Arteaga, and Alarcón, **81**

Chapter 6 Jamil Mahuad, **115**

Chapter 7 Levantamiento Indígena, **163**

Chapter 8 Gustavo Noboa, **205**

Conclusion, **235**

Notes, **249**

Bibliography, **263**

Index, **271**

Style Note

For the purpose of clarity, quotations have been italicized instead of placed within the usual quotation marks. Regarding translations from Spanish to English, some are my own, some are by other translators, and others have been adapted by me from translations already existing. When referring to locations, for the most part I have used the names of modern provinces, cities, towns, and so forth that were in use in the year 2000, even when contemporary names were not used during the time referred to in the text. It seemed simpler to keep the same names throughout the book rather than change them from period to period. Thus, for example, "Sucumbíos" and "Nueva Loja" refer to a province and a capital carved out of an older administrative district in the late twentieth century, although their earlier names were "Napo" and "Lago Agrio."

Abbreviations are a challenge. Ecuadoreans are fond of using initials for the names of political parties, social movements, Indian organizations, labor unions, government agencies, and a host of other groups. My practice is to write the full name of the entity at first mention, with its abbreviation immediately following in parentheses. Thereafter, in most chapters, only the initials are used. An example is the political party Nuevo País, New Country (NP), subsequently referred to as NP.

Introduction

In February 1997, Ecuador's President Abdalá Bucaram, widely known by a variety of nicknames including "El Loco," was removed by Congress on the grounds of mental incapacity. It seemed to his opponents the simplest way to depose the often wildly uninhibited leader. No fewer than three individuals simultaneously claimed the nation's top position, and the country proceeded to have three heads of state in as many days. For reasons of what it called "constitutional stability" the United States wanted one of them, Rosalía Arteaga, and for a few days it got her. She was rapidly painted into a corner, however, and told by the military she had to resign. Congress's head, Fabián Alarcón, then assumed the presidency until elections for new leaders were held.

Just three years later, in January 2000, yet another president, the calm and cerebral Jamil Mahuad, was toppled. Junior military officers in charge of guarding Congress and the Supreme Court simply stepped aside and allowed thousands of Indians, many of whom had come to town disguised in Cholo dress, to enter and occupy the buildings. The Indians and soldiers formed the so-called Junta of National Salvation and declared the old order abolished. With great prescience, the doomed Mahuad pledged he would never resign. Belligerently, but without a base of support, he announced that he would have to be removed by force. After that scenario was duly enacted, he appeared on television with a question-begging statement. *A thrown-out President does not resign,* he insisted. *He is thrown out.*[1] With two of three branches of government under their control, the Junta soldiers and thousands of Indians marched on Carondelet, the presidential palace, but only their leaders got inside. A few hours later their Junta was reconstituted when the head of the Armed Forces became a member, but in little time he abolished the body. The vice president proceeded to power, the soldiers went back to their barracks, and the Indians returned to the provinces. When Vice President Gustavo Noboa took the frayed reins of government in hand, his acceptance was a masterpiece of assumed modesty, followed with a one-sentence summary of his official muscle. *I find myself under the obligation to assume the presidency of Ecuador*, he noted, and added after a seemly pause: *I have the support of the Armed Forces and the National Police.*[2] Through it all the U.S. government warned that Ecuador's isolation would be akin to Cuba's unless the country followed the constitutional process, which, after a fashion, it did. The State Department's top official for Latin America, Peter Romero, described it in one word: *chaos*.[3]

On both occasions when presidents were toppled—Bucaram in 1997 and Mahuad in 2000—the leaders fell following massive demonstrations that blocked streets and highways and brought commerce to a virtual standstill. Each man was overthrown while Ecuador was in the midst of its worst economic crisis in over seven decades. Most of the population had suffered severely, but the nation's Indians, as always, were at the bottom of the heap. The popularity of each leader had plummeted soon after inauguration, and public opinion polls showed what the enormity of the protests against them demonstrated: over 90 percent of the people wanted them sacked.

The chaotic events at the end of the century followed eighteen years of orderly transitions of presidential power during which five democratically elected men served out their four-year terms, save one who was killed in a plane crash. After withdrawing to the barracks in 1979, the military stayed there for the better part of two decades, only to emerge to assume crucial roles during the Bucaram and Mahuad crises. What follows is an effort to describe and explain the tumult of 1997 and 2000. Neither collapse of government resulted in a radical upheaval as deep-seated as the revolutions in Mexico in 1910, Bolivia in 1952, or Cuba in 1959, although the January 2000 coalition of Indians and junior officers had the potential for fundamental change until it was thwarted by the military high command, traditional politicians, and the U.S. government. The following pages focus on the fall of two regimes but begin by placing them in the perspective of thirty years of dynamic change and transformation launched in 1967 by the discovery of oil in the Amazon. After the first barrel of crude was sent abroad five years later, forces were set in motion that continue to reshape Ecuador's social, economic, and political landscape.

This study reviews the main themes of the country's history during the last half millennium and concentrates on its evolution in the last third of the twentieth century. The closest focus is on developments from 1996 to 2000. The examination shows how increasingly interrelated are politics, economics, culture, the environment, finance, diplomacy, and more, not only within Ecuador but also on a global level. Any one of these topics cannot be grasped without looking at the others, and they must all be tied together for an understanding unobtainable by looking at only one factor.

The book is divided into eight parts: (1) the land and the people; (2) historical background, from the indigenous pre-Conquest past to 1532, Spanish colonial rule, and the 150 years from independence in the 1822 through 1972; (3) the oil era from 1972 through 1996; (4) the impact of oil in the Oriente and the emergence of a national Indian movement in the last half of the twentieth century; (5) the economic decline and political instability from 1996 to 2000, when Abdalá Bucaram, Rosalía Arteaga, and Fabián Alarcón vied for the presidency; (6) Jamil Mahuad and his administration; and (7) the Indian uprising, or Levantamiento Indígena, of January 21, 2000; and (8) its aftermath, with President Gustavo Noboa.

Ecuador's economic, social, and political past all receive a measure of attention, as many of its traditions continue to interact to mold the values, beliefs, institutions, and an array of practices that its people have developed. Many of them linger on, for better or worse, to influence the present and future. Among this study's main topics are the broad and deep impacts of oil; extraordinarily rapid, uneven, and inequitable economic development; so-called modernization; and dependency on foreign markets and international lenders. It also takes up deficits, burdensome foreign debt, the expansion and contraction of state spending and programs, rising popular demands, intense social conflict, and the emergence of the Indian movement. It deals as well with continued political fragmentation, near-constant corruption, the looming presence of the military as final arbiter in the nation's politics, and the powerful impact of U.S. policy on Ecuador's affairs. During the last three decades of the twentieth century, these themes took on a host of new characteristics and dimensions and were increasingly viewed through the lens of what came to be called globalization. Essentially, they explain much of the country's condition as one millennium ended and another began.

Petroleum transformed Ecuador. The wealth it generated fostered modernization, but it also wreaked havoc with the environment and cultures of the indigenous people of the Amazon. They fought back by forming federations to advance their interests and joined forces with similar groups in the highlands. Together they created Ecuador's first truly national indigenous organization in 1986, and by 2000 their movement was a major force to be reckoned with—one that increasingly influenced state policy and helped bring down two governments. The Indians battled for economic advancement, but above all they demanded respect for the dignity of their cultures and for their moral and historic rights to their lands and territories. They proudly reaffirmed their diverse coastal, highland, and Oriente origins and insisted upon equality as citizens of a plurinational and multicultural society.

Oil revenue made it possible to continue overdependency on one export, a tradition begun by cacao and continued with bananas, which maintained and, some say, heightened the nation's social, economic, and political vulnerability to fluctuations in the international price of a single product. During the oil boom, moreover, much of the population came to expect low taxes accompanied by high levels of state expenditures to provide social services. When the decline in petroleum prices and public revenue inevitably arrived, well-entrenched expectations and a concomitant reluctance to reduce outlays generated opposition to fiscal belt-tightening. Such measures were attempted by successive administrations, all of which struggled to balance budgets and maintain a modicum of social peace as strikes, protests, and demonstrations proliferated. Rather than risk too much austerity, politicians turned to deficit spending and foreign loans, financial approaches that fueled inflation. And, contrary to the expectations of many people, oil wealth did not end the old traditions of political fragmentation and corruption. Some claim it even worsened them, and both longstanding customs took on added significance during times of economic downswing. They helped bring masses of people into the streets and drive politicians to turmoil, all of which in turn prompted the military to dictate final arrangements. When matters strayed too far from the constitutional process, the U.S. government applied enormous pressure to help shape the ultimate political outcomes.

Contrasting Voices

The removal from office of President Jamil Mahuad in January 2000 provoked a wide range of comments, some angry, others thoughtful, all illuminating. The following statements shed light on where Ecuadoreans thought they stood before and after the event.

Comments beforehand:

One of the grave defects in our leaders has been to become excessively enthusiastic about sonorous government programs. The government is always directed by the nature of things, as Napoleon wisely said. Evil is not essentially in dictatorship, and good is not necessarily in democracy. Evil lies in applying oppression with perverse or vain intent. Good lies in making effective the rights of man and of the citizen and in the creation of institutions that guarantee them.

José María Velasco Ibarra, 1944. Velasco Ibarra was president of Ecuador five times between 1934 and 1972

We realistically accept the fact that only moderate reforms are possible in this country and we propose no more than that. We will work entirely within the existing framework.

Vice President Osvaldo Hurtado Larrea, 1979

In the future, the only way to arrive at a change will be through a revolutionary process. I am almost convinced of that.

President Abdalá Bucaram, March 1998

Popular support is important, but it is a historical constant in Ecuador that a president arrives, has to make tough decisions, and falls out of popular favor.

President Jamil Mahuad, April 20, 1999

Comments afterward:

The ruling class has destroyed the country. We've got to kill the entire political class and begin anew.

RIGOBERTO VILLARREAL, A 34-YEAR-OLD
QUITO TAXI DRIVER, JANUARY 23, 2000

Internal confidence is destroyed. The country's image is horrible. If we don't create jobs, if we don't generate production, if we don't stabilize the country's economy, Ecuador is finished.

MAYOR OF GUAYAQUIL AND FORMER PRESIDENT
LEÓN FEBRES CORDERO, JANUARY 23, 2000

It is indispensable that Ecuador has peace, but to have peace you need freedom, and to have freedom you need justice. And the Indian population needs justice.

PRESIDENT GUSTAVO NOBOA, JANUARY 23, 2000

We are going to continue the fighting. We may return to march on Quito and we could be a lot more hard-line when we mobilize again.

ANTONIO VARGAS, LEADER OF CONAIE, JANUARY 23, 2000

We are not going to cure the republic with witchcraft or with hostile protests against those who are not wearing ponchos.

INTERIOR MINISTER FRANCISCO HUERTA, GIVING HIS VIEW OF
TRADITIONAL PRACTICES OF THE INDIANS, JANUARY 23, 2000

The strong movement of sergeants and other officials is still there.

EDITORIAL IN *EL COMERCIO* NEWSPAPER, JANUARY 23, 2000

Ecuador is not a banana republic. We are going to modernize and change this country both politically and economically.

PRESIDENT GUSTAVO NOBOA, JANUARY 25, 2000

There is excessive fragmentation of parties and the president of the republic, lacking a majority in the parliament, dedicates a great deal of his energies to trying to reach agreement with the opposition. The fragmented politicians condition their agreements on their immediate electoral interests. It creates a climate of constant instability.

MILTON LUNA, DIRECTOR OF HISTORY AT THE CATHOLIC UNIVERSITY,
QUITO, WRITING IN *EL COMERCIO*, FEBRUARY 6, 2000

CONAIE is the major Indian organization on the continent. In contrast, Ecuador's political class is the least prestigious, weakest, and most backward in the Americas.

El Comercio COLUMNIST HERNÁN RAMOS BENALCÁZAR, APRIL 28, 2000

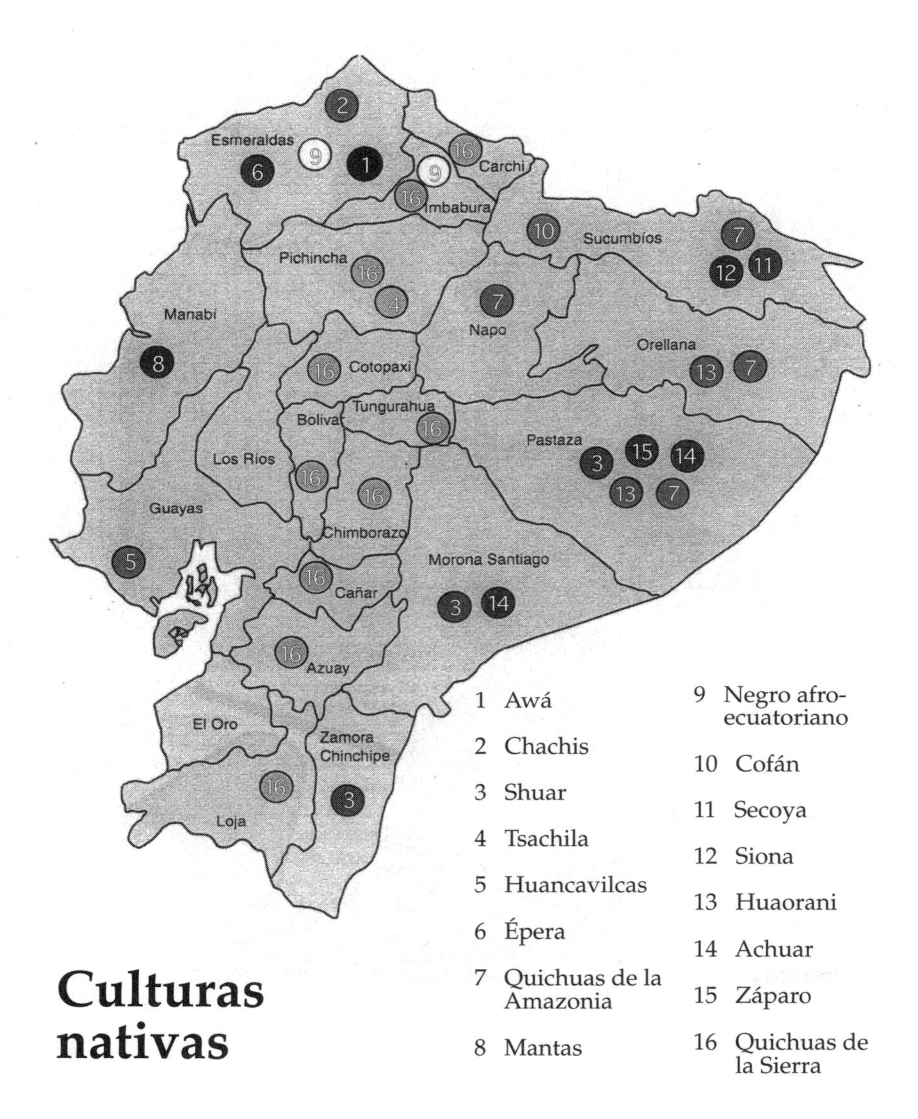

Culturas nativas

1

The Land and the People

Ecuador's human contrasts are striking, including those between rich and poor, city and country, learned and illiterate, lord and peasant, cosmopolitan entrepreneur and desperate street urchin. So, too, is its geographic diversity. The nation, which took its name in 1830 from the equatorial line that runs through it, is about half the size of France and a little smaller than the state of New Mexico. Located in northwestern South America, Ecuador's 108,623 square miles are bounded by Colombia on the north, Peru on the east and south, and the Pacific Ocean on the west. It includes the Galápagos Islands about 600 miles west of the mainland. The country's four main regions are the coastal plains along the Pacific (Costa), the snow-capped Andes of the interior highlands (Sierra), the Amazon jungle to the east (Oriente, a term that means the orient or east), and the Galápagos. Ecuador is organized administratively into twenty-one provinces; ten are in the highlands, while the coast and Oriente have five each, and the Galápagos Islands form another. Quito, the capital, lies in the highlands a mere four hours drive to the Pacific Ocean, three hours to the Oriente jungle, and two hours to mountain glaciers. Although Ecuador occupies only about 1.72 percent of the Earth's surface, it has an amazing diversity of flora and fauna for its size, including nearly 17 percent of the planet's species of birds.

For centuries following the Spanish conquest in the 1530s, natural barriers fostered a strong measure of isolation among regions, and diverse economic and social systems evolved with frequently conflicting and competing interests. Nature's principal obstacles were the steep Andean mountains in the interior, which separated the bulk of the people in the highlands from those on the coast, and the thick vegetation of the Amazon rain forest, which for centuries made the Oriente almost impenetrable to outsiders. Ecuador's development was also influenced by its being the only Andean country without abundant precious minerals, although oil was found in the Amazon in the middle of the twentieth century. Agriculture and textiles dominated the economy of the highlands for centuries, where production focused on corn, potatoes, wheat, cattle, pigs, sheep, and wool and cotton textiles. The society was highly stratified and controlled by a small group of white landowners whose ample estates were worked by a large underclass of indigenous people.

The coast constitutes one fourth of the country, runs the length of the Pacific for 350 miles, and varies in width from 12 to 112 miles between the ocean and the Andes. Most of it is under 1,000 feet above sea level, although a chain of hills rises in the north to around 2,500 feet. Dozens of rivers that flow from the snow-capped mountains to the Pacific cut the coast and often overflow to cause destructive floods during the rainy season from January through April. The Esmeraldas River to the north and the Guayas and Daule to the south are the principal water systems. Coconut palms line much of the shore, whose estuaries harbor mangroves that grow in saltwater swamps. Plantations of bananas, cacao, sugarcane, and rice are farther inland, particularly in the fertile irrigated lowlands. This area nurtured South America's first crop of bananas, and in 2000 Ecuador was the world's top producer of the fruit. Coastal agriculture and fishing continue to be directed toward export, in contrast to the focus on domestic consumption of farmers in the highlands. The republic's principal port, through which most exports and imports have always flowed, is Santiago de Guayaquil, thirty miles inland from the Pacific on the bank of the Guayas River, a body formed by the union of the Daule and Babahoyo. Guayaquil was the nation's largest urban center in 2000; its centuries-old commercial and business prominence came from its location at the center of coastal agriculture.

Like the coast, the highlands region makes up one-quarter of the country. It contains over one-half of the people, including about 96 percent of the indigenous population, and runs 390 miles long and a maximum of 45 miles wide. The huge, rugged Andes consist of two parallel north-south mountain chains; the eastern one is called the Cordillera Oriental, and the western the Cordillera Occidental. Although the eastern is higher and older, the western has the highest peak, Mount Chimborazo. Between the Cordilleras is an enclosed and narrow plateau, about 7,000 to 9,500 feet in elevation and interrupted at intervals by foothills and high grasslands called *páramos*. They run across the plateau, roughly east to west, and connect with the double row of massive mountains. Nestled inside the mountains and *páramos* lie a series of enclosed and fertile valleys (*hoyas*) with at least one river running through each. Quito is in the Guayllabamba valley; its river is the Machangara. In the ten most prominent valleys, from north to south, are situated Tulcán, Ibarra, Quito, Latacunga, Ambato, Guaranda, Riobamba, Azogues, Cuenca, and Loja. They are the capitals of the ten highland provinces.

These valleys are home to the bulk of the indigenous population, including, among others, and again from north to south, the Caranqui, Otavaleno, Cayambi, Quitu, Panzaleo, Chibuelo, Salasacan, Tugua, Waranka, Puruhá, Cañari, and Saraguro. The area resembles a ladder, with the *páramos* making up its rungs and the double row of north-south mountains forming the sides. Narrow passes penetrate the rungs and connect the valleys. In these populated and well-farmed valleys the Spaniards encountered the indigenous population, conquered by the Incas only a few decades earlier, and there they made their main settlements. Adjoining small parcels of land are cultivated at different times of the year, depending on what crop is sown. Wheat and barley together, then potatoes, then beans and corn, and so forth, making the fields a dynamic landscape of diverse colors and rare beauty. The Andean condor, the largest flying bird, inhabits the high *páramos*, particularly near the heights of Tungurahua, Cayambe, Antizana, and Altar. One finds mountain tapirs, bears, pumas, deer, and more. Domesticated cattle, pigs, sheep, and horses were all brought in by the Spaniards; and during the nineteenth century eucalyptus trees from Australia, whose seeds grow well in poor as well as rich soil, were planted the length and breadth of the highlands to combat soil erosion.

Ecuador's Andes have three major regions. The north, including Carchi, Imbabura, Pichincha, and Cotopaxi provinces, has the most fertile farmland and the largest extensions of grassland. The center is similar, composed of Tungurahua, Chimborazo, and Bolívar provinces, but it has smaller agricultural and livestock zones. The south is less fertile, with more volcanic soil and less advanced agricultural development, and is made up of Cañar, Azuay, and Loja provinces. Quito lies at 9,350 feet in a northern valley at the base of a 17,716-feet high and occasionally active volcano called Pichincha. Many of the volcanoes of the central highlands have erupted over the centuries, and many still remain active. On October 7, 1999, for example, Pichincha spewed enough white ash over Quito to close the airport for a day and foster brisk sales of paper mouth masks. In early 2000, Tungurahua threatened to erupt for the first time since 1918. In Baños de Agua Santa, near the base of the mountain, 1,800 businesses closed as a consequence and the town was almost evacuated. The highland valleys and the capital are linked by the Pan American Highway, whose economic and political significance in moving produce to urban markets was by 2000 well understood, as members of the indigenous movement frequently blocked it to protest government policies.

The Oriente, also referred to simply as the Amazon, makes up about half of the country and lies east of the Andes. The Amazon River springs from the eastern cordillera's high glacial regions, and the riddle about its source inspired speculation for centuries. In December 2000 a five-nation expedition determined its origin through the use of orbiting satellites. The river starts as a stream on Nevado Mismi, a summit in southern Peru. The Oriente has what some call Ecuador's third Andean mountain chain, a broken series including the Napo Galeras, Cutucú, and Cóndor ranges, which rise to 4,000 feet before descending eastward and which includes cone-shaped Reventador in northern Napo province. Rain and melting snow high in Ecuador's Andes flow east, ultimately to help form the Amazon, and the country's 530-mile-long Napo River in the north is the country's longest affluent leading to it. Other major rivers include the Putumayo, San Miguel, Aguarico, Coca, Pastaza, and Santiago. The Amazon jungle pulsates with leaf-cutting ants, cicadas, and other insects; butterflies and boas, toucans, and a multitude of brightly colored birds, frogs, and snakes; a host of reptiles, tapirs, monkeys, and more. It has a multitude of broad rivers, waterfalls, and striking vegetation, including a wide variety of orchids. Long home to the Cofán, Secoya, Siona, Huaorani, Achuar, Záparo, Shuar, and Oriente Quichuas, the once near-isolated area underwent profound change in the 1960s with the incursion of outsiders, including those engaged in oil, timber, agriculture, and mining activities. Centuries-old cultures were deeply altered by the infusion of outsiders, and much of the rain forest's animal and plant life was diminished or destroyed along with the information it doubtless contained.

For most of the twentieth century, to many outsiders the Ecuadorean Amazon was strongly associated with the second largest nationality in the region whom they called the Jívaros, a group who always called themselves the Shuar. Inhabiting Pastaza, Morona Santiago, and Zamora Chinchipe provinces in the center and south, they were thought of as killers who cut off the heads of their enemies and shrunk them into trophies. The practice has been outlawed. The Amazon region has likewise come to be viewed differently, primarily as one whose enormous rain forest is appreciated as one of the world's vital oxygen suppliers that is being slowly depleted by modern man.

The Galápagos are named after the giant tortoises that inhabit them, which may live up to 150 or 200 years. The islands are volcanic in origin, formed about 15 million years ago by solidified lava that erupted from craters some 30,000 feet from the surface on the ocean floor. They have level shorelines and mountainous interiors with craters, some of which rise to 5,000 feet and a small minority of which remain active. The archipelago includes six large, nine small, and several hundred tiny islands on or near the equator.

The islands were uninhabited when the Spaniards first explored them in 1535, and during the seventeenth and eighteenth centuries they were frequented by pirates and buccaneers. British and U.S. warships and whaling vessels stopped at the archipelago in the nineteenth century, but it was not settled until annexation by Ecuador in 1832. Three years later, 26-year-old Charles Darwin spent six weeks on the islands studying plant and animal life, and his observations furnished vital data for his brilliant explanation of the process of evolution, *Origin of Species* (1859). The Galápagos are still noted for their animal life, including the giant tortoises, land and marine iguanas, sea lions, seals, penguins, pelicans, blue- and red-footed boobies, frigate birds, flamengos, finches, and a plethora of other birds. Tourists, fishermen, and new species introduced mostly by accident, however, have so altered the Galápagos that in 1997 an emergency decree was issued to protect the natural preserve, which the United Nations in 1979 officially designated a "Cultural Patrimony of Humanity." In 1999 an estimated 66,000 tourists visited the islands; about 81 percent of them were foreigners, and the vast majority were environmental rather than beach tourists. In 1950 the islands had 1,500 inhabitants, a number that grew to 15,300 in 1998 and climbed to 17,000 by 2000. Most were concentrated in about 3 percent of the territory, and the majority were fishermen.

Ecuador's climate varies by region, but in each zone the only two seasons are dry summer and wet winter. In the highlands the dry summer spans June through September and comes again for one month in late December and early January; April is the wettest month of the winter. On the coast and in the Galápagos the dry summer runs from May to December; the rest is wet winter. In the Oriente the driest time is from September to December, but it rains year round with most precipitation descending from June to August. The economic and other effects of El Niño have long been familiar to Ecuadoreans, as the impact of the weather pattern is particularly strong on the coasts of Ecuador and Peru. When it occurs in the Pacific, unusually warm ocean water appears along the western coast, causing the climatic disturbances of unpredictably strong winds and rains. As elsewhere in the world, this aspect of what was once simply called "weather" is now being studied as a classic example of the "chaotic systems" that frequently defy timely intervention and that often take an almost unbearable toll in human misery.

Ecuador's people and cultures are as diverse as its geography. The indigenous population descends from pre-Colombian inhabitants, who ever since the Spanish conquest have occupied the bottom rung of the social and economic ladder. The progeny of the Spanish conquistadores and settlers, though far fewer than the natives, stood at the top of society as commanders of most of its technology, weaponry, wealth, and power. Those whose ancestors were brought as slaves from Africa are the smallest group as well as among the poorest, while the mixture of Indians and Spaniards (Cholos or Mestizos) are the largest, occupying the middle social and economic levels.

In the year 2000 there were strong indications that Ecuador's long-suppressed, exploited, and manipulated Indians were moving from relative political isolation toward center stage and in the process were broadening and deepening the imagined Ecuadorean community. Their accomplishment had taken centuries of struggle. It was still uncertain, however, whether a pluralistic, multicultural society would be established and accepted, and whether unity would be achieved through diversity or through the fusion of vastly different worlds and cultures. What kind of society would be molded, and for the betterment of whom, remain unanswered questions. Whatever may be the answers, modern-day Ecuador and its historical development, and the events from 1996 to 2000 that are the focus of this study, cannot be understood without addressing the evolution of the different ethnic groups struggling to live together.

The national census located 4,581,476 Ecuadoreans in 1962 and 10,800,000 in 1991. The population doubled in three decades. In 2000 the number stood at 12.4 million, organized in 2,476,267 families, and was growing at the rate of 2.4 percent per year. The pace suggested another doubling in three decades. Most were young, with 40 percent under age fifteen, compared to 22 percent in the United States and 19 percent in Britain in the same age group. A little over half, 55 percent, were classified as urban; the rest were labeled rural. Ecuador's population growth was similar to the rest of Latin America, which rose from 150 million in 1950 to 436 million by 1990, a 290 percent leap. The rapid growth was part of a relatively recent trend worldwide. What set the numbers soaring was the twentieth-century public health revolution, with vaccines and antibiotics that seemed to come almost all at once and that rapidly spread around the globe at midcentury. Equally dramatic was the revolution in food production, which tripled during the last half of the twentieth century and far outpaced population growth. In the United States alone, if people had died throughout the twentieth century at the same rate as they did at its beginning, the population now would be 140 million, not 250 million. United Nations analysts predict that human beings will about double in number one more time, reach about 10 to 11 billion, and then stabilize and decline.

Efforts to identify and count on a racial basis are both approximate and controversial. Some claim the Ecuadorean population is 40 percent Indian, 40 percent Cholo, 10 percent white, and 10 percent black. Others insist the composition is 25 percent Indian and 55 percent Cholo, a mixture of Indian and white, while many simply say that about 80 percent of the people are Indians and Cholos, with the remainder equally divided between blacks and whites. On May 7, 2000, with Ecuador's population at about 12.4 million, Quito's newspaper *El Comercio* aptly summarized the question of racial classification:

> To establish the number of Indians that exist in Ecuador is a theme that has occasioned more than a headache. The problem begins when one tries to establish who the Indians are, which is to say, does one consider them such on the basis of physical characteristics or whether they live within the Andean Indian world? White and Mestizo society have affirmed—among them former presidents León Febres Cordero and Osvaldo Hurtado—that the Indian population is a little less than two million people and that, as such, it has no right to impose its vision of the world on Ecuador. The organized Indian movement, on the other hand, says its population numbers over four million people.[1]

What is not in dispute is that in the early sixteenth century Spain conquered the area that four centuries later came to be called Ecuador, that the Conquest imposed a new social order characterized by racially based domination, exploitation, and hierarchy, and that an intermingling of European, African, and indigenous people followed. Both conquerors and conquered were transformed in the process, and the racial and ethnic groups involved still struggle to find mutually acceptable terms and perceptions in which to coexist.

Just as one can only surmise the number of Ecuador's indigenous people in 2000, so, too, is it mostly a guess that one million of their ancestors lived in the area when Sebastián de Benalcázar arrived in 1534. Some analysts, including the Confederation of Indigenous

Nationalities of Ecuador (CONAIE), set the Indian population in 2000 at one-third of the total, about 4 million of Ecuador's 12.4 million inhabitants. Eleven years earlier, however, CONAIE's estimate was 7.5 percent lower; in 1989 it claimed that the country had 3,111,900 Indians. About 96.4 percent or 3 million of them were highland Quichuas, the confederation reported. Living along the coast were 0.24 percent, or 7,600 natives, among them 1,600 Awá, 4,000 Chachi, and 2,000 Tsachila. The Oriente was home to 3.35 percent, or about 104,300 Indians, including 60,000 Quichua, 40,000 Shuar, 2,000 Huaorani, 1,000 Siona-Secoya, 800 Cofán, and 500 Achuar. If CONAIE's figures were correct, as the country's population in 1990 was 9,648,189 the indigenous population constituted 32.25 percent of the total. Many disputed CONAIE's estimates and argued that both the 40 and 32.35 percent figures were too high. Their reasoning was that the coast contained half of all inhabitants and was home to very few Indians, and that the same was true of the Oriente; indeed, there were scarcely more than 100,000 Indians in the two regions. They believed the majority of highland people were Cholos and concluded that the natives were under one-quarter of the total population of 12.4 million in 2000. Noted historian Enrique Ayala Mora argued that at the millennium the indigenous population was 16 percent at most and constituted a majority in only two highland provinces, Chimborazo and Bolívar.

In the highlands the majority of Indians are Quichua-speaking descendants of subjects of the Incas who emerged about 1100 AD near Cuzco, Peru, and who brought the area of modern Ecuador under their control in the late fifteenth and early sixteenth centuries. In 2000, linguistic variances lingered between provinces, a reflection of centuries of isolation imposed by the rugged terrain of the Andes. They were also the result of strong community ties that inhibited movement, of labor laws that restricted mobility, and of the fact that the Inca invaders had arrived and imposed their language only a few decades before Benalcázar conquered Quito. Quichua speakers live in all highland provinces and include, from north to south, the Caranqui and Otavalenos of Imbabura; the Cayambi of Imbabura and Pichincha; the Quitu of Pichincha; the Panzaleo, Chibueleo, and Salasacan of Tungurahua and Cotopaxi; the Tugua of Cotopaxi; the Waranka of Bolívar; the Puruhá of Chimborazo, El Oro, and Guayas; the Cañari of Cañar and Azuay; and the Saraguro of Loja (many of whom have emigrated to Shuar territory in Zamora-Chinchipe province). Chimborazo has the highest percentage of highland Indians, while Carchi province along the border with Colombia has the fewest. Highland Indians are marked by a number of differences, above all in their allegiance to their respective nationalities and communities, and are distinguished by local traditions and dress and linguistic variations. The Otavalo people of Imbabura are known worldwide as superb wool weavers; the men wear traditional blue ponchos and white pants, colors worn by women as well. The Salasacans of Tungurahua use black and white ponchos along with white hats with broad upturned brims. Likely the Incas forced them to move from the southern part of the empire, as their Quichua includes a number of terms common to Bolivia but not spoken by other indigenous groups in Ecuador. The Saraguros of Loja are likewise easy to recognize, as they dress in black ponchos made from the wool of black sheep. Many analysts believe that the Incas moved the Salasacans and Saraguros from Bolivia to Ecuador as *mitimaes*, people shifted from a secure part of the empire to help incorporate a newly conquered region.

Natives of the Amazon area have always numbered far fewer than their counterparts in the Andes. Over the centuries they evolved a sophisticated adaptation to their jungle environment, one characterized by highly complex interaction with an enormous array of plants and animals. Dye from achiote extracted from annatto seeds, for example, continues to decorate faces and extremities, adorn pottery and clothing, keep insects away, and ward off the common jungle problem of foot fungus. Extracting poisons from select plants, to further make the point, made the tips of darts launched by blowguns effective against monkeys and other tree-dwelling animals, as did partially cutting the darts a little below the end with the teeth of piranha fish to help ensure they would break off and remain long enough inside the body to paralyze the prey. And over time the area's natives learned to extract poisonous prussic acid from a bitter type of manioc, which, when made into biscuits, both lasted and nourished exceptionally well. Nature provided other levels of support for the rain forest's indigenous people. Waterfall rites and reverence to trees, boas, jaguars and other creatures helped to spiritually ground the Indians in the cosmos.

The most numerous of the Amazon region's Indians are related to the highland Quichuas. They are called the Oriente Quichuas and are divided into the Canelos and Quijos of north-central Sucumbíos, Napo, Orellana, and Pastaza provinces. Both groups make exceptional pottery painted with red, black, and white symbolic drawings of jaguars, boas, frogs, and other jungle animals. Their delicate paintbrushes are made of human hair, and, like all Oriente pottery, their pots, jars, bowls, and dishes are made by hand coiling instead of on the potter's wheel.

The second largest group are the Shuar, closely related to the Achuar, who inhabit the central-south provinces of Pastaza, Morona Santiago, and Zamora Chinchipe. Like people everywhere, they developed myths to explain life in general as well as the interaction of various forces around them and what they deemed appropriate responses to them. The Shuar were traditionally fierce warriors, and after killing an enemy they would cut off and shrink the victim's head. (The reduction was simple but time consuming. The skull was removed before the beginning of the process, which entailed scraping the skin clean and heating it in hot water for no more than thirty minutes lest the hair fall out, and burying it in the ground until it dried and turned leather-like. This procedure reduced the trophy's size by half. It was then sewn and stuffed with hot sand, a step repeated several times. A red-hot machete dried the lips, and the whole was rubbed with charcoal.) For the Shuar, head or *tsansta,* hunting was a way to accumulate protective spiritual powers referred to as *arutam*. Carrying the *arutam tsantsa* was necessary to keep the victim's soul from avenging his death. Those in possession of many shrunken heads were spared their own murder or the anxiety of meeting misfortune. A warrior was not born with *arutams* but had to obtain them repeatedly, as their power lasted only four to five years. Today the Shuar live in ways radically different from their ancestors and are often involved in national politics.

The Huaorani, whose language is unrelated to that of any other Oriente group, occupy the center provinces of Orellana and Pastaza. Men's traditional dress included only a cord made of wild cotton wrapped around the waist to keep the penis upright. They felt naked and defenseless without it, as it represented power and energy central to their culture. Polygamy was a common practice. Excellent hunters who used blowguns to send poison-tipped darts into howler monkeys and other animals, the Huaorani engaged in near-constant warfare over the centuries and were historically called Aucas or "savages," by the Quichuas. Their traditional lands have abundant oil, and their lives changed dramatically when it was found in 1967.

Orellana and Pastaza are also home to the Achuar, who, more than most Indians, prized their independence and resented outsiders. Achuar warriors painted their faces with achiote, taking special care in drawing around the mouth to make it resemble a jaguar ready to kill. Women cultivated manioc, from which a fermented drink was made, as well as plantains and a variety of other vegetables and fruits, while men hunted with long, slender blowguns. Like many other indigenous groups in the Oriente, the Achuar induced hallucinogenic trances from substances found in jungle plants as a way to gain *arutam* strength and courage as well as to enter what were deemed higher realms of existence.

The Cofán live in the western portion of Sucumbíos province that borders Colombia, while the Siona and Secoya inhabit its eastern portion. All three traditionally dressed in tunics of barkcloth, made somewhat soft by repeated washing and pounding. Each produced pottery using the coil technique and made canoes through a system of reciprocal communal labor called the *minga*. The Siona and Secoya were traditional enemies of the Huaorani to the south, with whom they fought frequent wars, but during the second half of the twentieth century they faced a different sort of foe. Their way of life underwent enormous pressure and major change from oil companies that ended their relative isolation.

The smallest group of Oriente Indians are the Záparos of Pastaza province in the region's center. Like the clothing of the Cofán, Siona, and Secoya, the traditional dress of the Záparos was made from tree bark and was commonly decorated with achiote dye. On ceremonial occasions they used paste from another seed to dye their hair intensely black and to paint designs on their faces and extremities. They were skilled in the use of blowguns and lances. The Záparos almost disappeared due to epidemics of measles, smallpox, and influenza. Other nationalities that did die out include the Arda, Bolona, Bracamoro Chirino, and Tete.

For centuries the Amazon dwellers differed in fundamental ways from highland natives who rarely strayed from their place of birth. Traditional farming, for example, involved clearing a few acres of the ecosystem's poor soil every few years for a new field of manioc, sweet potatoes, peanuts, and other crops that flourished with very little care. The jungle simply took over the fallow plots left behind. This practice, along with hunting and fishing, required a substantial amount of land, although the borders of territories between different indigenous nationalities were marked and recognized. Particularly after the 1960s, however, ancient agricultural and cultural practices changed dramatically as the number of outsiders entering the area multiplied, among them missionaries, colonists, lumbermen, and oil workers. Cattle, fruit trees, and numerous new products and practices were introduced, all of which diminished the amount of available land and altered the traditional ways of life. Modern medical treatment and drugs, provided primarily by missionaries, reduced mortality rates, especially among infants. In the 1950s, for example, the Shuar likely numbered about 7,800; in the 1970s, estimates ranged over 15,000, and in 1989 the figure was put at 40,000 with a prognosis of sustained growth. Greater contact with the outside world led to increased political and cultural assertion, an early manifestation of which was the rejection of insulting names given the natives by outsiders.

With the exception of the Galápagos, the coast is the region with the fewest number of Indians. Three ethnic groups dominate. The Awá live in the northern coastal province of Esmeraldas, as do the Chachis, referred to by some as the Cayapa, and the 150 or so Épera. The area is extremely humid and supports a wide array of plant life, and offshore waters have abundant shrimp, oysters, clams, lobsters, and fish. The Tsachilas are found around Santo Domingo in the tropical zone of Pichincha province, and until recently were called Colorados because of the red achiote paste that many wore in their hair to guard against both insects and rain. The Mantas are in Manabí, and the Huancavilcas inhabit Guayas. The three provinces with the fewest indigenous people are Los Ríos, El Oro, and the Galápagos.

Decades before the discovery of oil impacted the Indians of the Oriente, the cultures of many coastal natives were substantially altered by job opportunities on the region's vast banana, coffee, and palm nut plantations and by influences from highland and Colombian migrants. Some Indians, however, sought a measure of control over the forces changing their lives. Several thousand Awá live in Colombia near the border with Ecuador, for example, and in 1986 the two countries established a reservation, which straddles the frontier, to include all of them. Within a 400-square-mile territory the natives were given preferential rights to land and natural resources along with the responsibility for their preservation and management.

Galo Plaza Lasso, president for a time in the early 1950s, said of his countrymen that *we are all coffee and milk, some more coffee than milk, others more milk than coffee, but nothing more.*[2] In 2000, well over half the people considered themselves Cholos or Mestizos, descendants of unions between whites and Indians. The term Cholo is more common than Mestizo in Ecuador, although both are heard, and it is usually used as a term of endearment. While the names have racial overtones, they also convey social meaning. An Indian, for example, who wears western dress and speaks Spanish is likely to be called a Cholo or Mestizo and may consider himself one, all of which contributes enormously to the challenge faced by those interested in dividing and tabulating the population on the basis of race or ethnicity.

The so-called whites, like the Cholos or Mestizos, speak Spanish and dress like Europeans. For centuries considerable prestige was attached to light skin color, and in relation to their numbers whites held a disproportionate share of power and wealth. For centuries the comparatively small number of wealthy whites excluded the vast majority of Indians from full participation in their imposed European concept of society and the nation. While they brought many western ways, they also established and long allowed to continue the exploitative legacies of colonialism and underdevelopment and did little to integrate the nation or foster fairness among its members of multiple cultures. In 2000, if whites were 10 percent of the population, as many observers claimed, then they numbered about 1.25 million.

Afro-Ecuadorean organizations estimated the country's black population at around 800,000 in 2000. CONAIE placed the figure at 500,000, while Ecuador's Center for the Study of Population and Social Development put the number at 55,800, or 4.5 percent of the total, demonstrating again how difficult and controversial it is to categorize people by race. Most blacks live on the coast, and many work on banana plantations or in other agricultural spheres. The greatest concentration of black communities is in the north, in Esmeraldas on the northwest coast and in Imbabura province in the northern highlands. A Spanish frigate carrying a cargo of slaves from Panama to Peru shipwrecked in the sixteenth century in Esmeraldas, where their descendants remain. The other largely Afro-Ecuadorean area is in Imbabura, in the warm, tropical highland Chota Valley. Chota is inhabited by descendants of slaves brought during the colonial period to work on Jesuit sugarcane plantations.

The music of those in Chota is a mixture of highland Indian instruments and rhythms and those of Africa, and soccer is as frequently and fervently played in the valley by blacks as it is by whites and Mestizos throughout the country. Yet the acceptance of Afro-Ecuadoreans and an end to discrimination against them have been slow. Blacks were often denigrated and condescendingly called "Negritos." By 2000, however, there were hopeful signs that such attitudes were changing, in large part owing to the efforts of Afro-Ecuadorean organizations, to the Indian movement's emphasis on racial equality and social justice, and to civil rights movements and the worldwide cultural assertion of blacks. In March 1999 the first congress of blacks was held, and from it emerged the National Confederation of Afro-Ecuadoreans, which formed a number of committees to focus on health, education, violence, political participation, women, ethnicity, culture, and history. The confederation established itself in the coastal provinces of Esmeraldas, Guayas, and El Oro; the highland provinces of Carchi, Imbabura, and Pichincha; and in the Oriente provinces of Sucumbíos and Orellana.

2

Historical Background to 1972

Ecuador's past strongly conditions its present. Understanding the kind of society in which the present generation lives requires an appreciation of the region's continuities and changes over the last five centuries. This brief survey emphasizes the indigenous past.

Pre-Inca cultures in Ecuador left little in the way of recorded history, and although archaeological evidence of their civilizations is growing, vast parts of the country remain largely uncharted. The conquest by the Incas of the numerous and diverse indigenous communities of Ecuador began around 1463 and, after fierce resistance by local tribes, by 1500 the highlands were under their control.

At its peak, the Inca empire stretched for almost 3,000 miles along the Andes from central Chile to the southern tip of Colombia. Ecuadorean peoples were a part of it—a complex, agriculture-based theocracy of sorts, well organized and well run. The emperor's title was Sapa Inca, and he was considered divine, a notion that confirmed absolutely his legitimacy. His close relatives held the top administrative positions, and their rule was backed by military forces stationed in permanent garrisons throughout an empire of four major regions, eighty provinces, and over 160 districts. The smallest unit was the agrarian community called the ayllu. New subjects were often moved to older population centers to encourage assimilation. A vast system of roads and rope bridges suspended over rivers and canyons facilitated rapid communication through the empire by message-carrying runners, and multicolored knotted strings called quipus served as sophisticated numerical and memory aids for a multitude of administrative tasks.

Religion was at the core of Inca society. Rulers claimed descent from Inti (the sun god), and instituted the practice of ancestor worship; the mummified corpses of emperors were carefully maintained and brought out during important ritual celebrations. For the common people, sacred items, persons, or places called huacas were central to religious life, and each ayllu had several of its own. While the Incas spread their religion throughout the empire, they tolerated local cults and practices.

Land was held collectively. Leaders called curacas headed the ayllu, and each family in the unit received a small plot on which to grow their own food. In addition, ayllu members were obligated to work lands for the support of the bureaucracy, military, and clergy. Government officials supervised crop selection and planting, taught irrigation, drainage, terracing, and fertilizing techniques, and ensured that a portion of the harvest was stored in warehouses for use during times of crop failure. The individual was firmly kept in second place to the collective. The only large domesticated animals were the llamas and alpacas. Neither were powerful beasts and served only as light pack animals and sources of meat and wool. Inca accomplishments were both numerous and impressive. To highlight but a few, their masons excelled in stone work, their artisans in textiles, ceramics, silver and gold work, and their surgeons performed skull and brain operations.

Ecuador rapidly gained importance within the empire. Sapa Inca Huayna Cápac settled in Quito, where he considered establishing a second capital and kept a large army and bureaucracy. Huáscar, the eldest of his sons, was fathered in the fashion of royal incest from a union with the Inca's sister in Cuzco (now Peru). Another son, Atahualpa, was the offspring of Huayna Cápac's marriage to a lesser wife in Ecuador. In 1527, Huayna Cápac died suddenly from a strange illness that felled thousands of natives in the mid-1520s who were without immunities against the deadly disease, perhaps smallpox or malaria, introduced from Europe, which could have spread along indigenous trade routes from Panama through Colombia and on into Ecuador.

After Huayna Cápac's sudden death, a ferocious five-year civil war broke out between his two sons, who were half-brothers, for control of the empire. Atahualpa had the professional army in the north that his father had used to bring Ecuador into the empire. When Huáscar's forces from Cuzco in the south tried to invade Quito, they were beaten back and defeated on the outskirts of Cuzco. This bitter conflict over the dynastic inheritance left the Incas weakened and divided at the moment when they desperately needed unity. Had the Spaniards arrived only a few years earlier or later, they would have encountered a cohesive empire. But in 1526 they faced an empire in crisis and were aided by thousands of Indians who either resented Inca domination or had recently sided with Huáscar in the civil war.

When the Spaniards arrived in the Americas, they were medieval men molded by the legacy of Rome and Christianity, and their centuries-long struggle to recapture the Iberian peninsula from Muslims left them with a religious militancy and creed that lacked respect for any cultural differences. When they landed in northern Ecuador in 1526, the conquistadores brought diseases, steel weapons, and horses, all of which were new to the natives. Six years later, Francisco Pizarro marched from the coast to the highlands with a force of sixty-two horsemen and 106 footsoldiers to begin his domination of Tawantinsuyu. Although he faced an empire of millions, in all but numbers the odds were against the Indians. The Incas had a small amount of bronze used for weapons and other objects, but besides alloying copper and tin they had not developed iron or other hard metals. Without iron, horses, and gunpowder, the Indians faced mounted Spaniards who wore metal armor, charged with rigid iron swords and spears, and fired guns and artillery.

In devising his plan to capture Atahualpa, Pizarro had in mind the same basic tactic that Hernán Cortés had used earlier in Mexico. He wanted to capture the divine head of state with the expectation that the masses then would capitulate and follow the orders of their imprisoned leader. The plan worked well. The assembled Inca army numbered at least 40,000 when in November 1532 Atahualpa met Pizarro at Cajamarca in northern Peru. Likely, Atahualpa was confident of the strength of his huge nearby army, and clearly he was intrigued by the Spaniards and their horses. The encounter did not turn out as he expected. Mounted Spaniards emerged from behind buildings in the city to slaughter some 6,000 to 7,000 natives in a surprise attack that lasted about two hours. The Indians, armed with only small axes, slings, and pouches of stones, could not retreat because of the small entryway into the city.

During the carnage the Spaniards captured Atahualpa, and later demanded a huge ransom and the dispersal of his troops in return for his release. His subjects deemed him a god and remained largely passive, just as Pizarro had calculated. Yet many of his soldiers immediately returned to Quito where one of Atahualpa's generals, Rumiñahui, had established himself as regional military ruler. Fearing Huáscar more than the Spaniards, Atahualpa ordered his half-brother killed rather than have him lead the resistance against the Europeans. But once the ransom gold was collected, Atahualpa was tried and executed on charges that included the murder of Huáscar. With both Sapa Inca brothers dead, the Incas were leaderless and demoralized, although Rumiñahui and others resisted Spanish advances.

Rivalry vexed the Spaniards. One force arrived on the coast in early 1534, led by Pedro de Alvarado from Guatemala, who was lured into the area by reports of Atahualpa's ransom and vast amounts of gold in the Andes. Sebastián de Benalcázar, nominally under the command of Pizarro, led the rival contingency. Benalcázar moved rapidly on Quito. Pizarro's partner, Diego de Almagro, then headed north from Cuzco to both thwart Alvarado and determine Benalcázar's intentions.

Meanwhile, Rumiñahui controlled Quito after the executions of Atahualpa and his brother Quillisacha and sent his lieutenant, Chiaquitinta, south to attack Benalcázar. The Indian forces fled upon encountering a Spanish reconnaissance unit mounted on thirty dreaded horses, and Benalcázar continued north to Quito. About 3,000 Cañaris, who viewed the Spaniards as liberators from their recent and brutal domination by the Incas, came to his aid and greatly eased his advance. Benalcázar finally reached Quito on June 22, 1534. It had taken him four months to get there, owing to stiff indigenous resistance. Rather than leave the city intact for the invaders, Rumiñahui evacuated it and razed it five days before the Spaniards entered Quito. Benalcázar proceeded to found the capital anew on December 6.

Benalcázar dispatched his lieutenant, Rui Díaz, to pursue Rumiñahui, who assembled 15,000 warriors and launched a night attack against the Spaniards who remained in the capital. After Rumiñahui's soldiers penetrated the city, repeated charges by Spaniards on horseback forced Rumiñahui to flee, and the following day seven local Indian leaders came to Quito to surrender to the new conquerors. Meanwhile, Atahualpa's general, Quisquis, who had killed most of Huáscar's family to eliminate any rivals to the throne, moved into southern Ecuador. Quisquis was killed by his own mutinous troops who were defeated in their first confrontation with Benalcázar's. Once again the primitive weapons of the Incas proved ineffective against Spanish horsemen in armor equipped with guns and iron lances. When Alvarado arrived in Quito in August 1534, Almagro, backed by Benalcázar, was in control of the city. Although both camps readied for battle, Almagro avoided conflict when he bought out Alvarado's ships and equipment for 100,000 gold pesos. Rumiñahui was pursued, captured, and burned alive in Quito's central square. He had orchestrated the most determined resistance faced by the Spaniards.

Most Spaniards viewed the Americas as territory to be conquered and its inhabitants as objects to exploit and convert to Christianity. Yet almost from the outset of their contact with the natives, the Spaniards were deeply divided in their opinions concerning the nature of the Indians. They tended to look upon them in extreme terms. The sixteenth-century Franciscan friar Bartolomé de las Casas, for example, wrote that *God created these simple people without evil and without guile. . . . They are most submissive, patient, peaceful and virtuous. Nor are they quarrelsome, rancorous, querulous, or vengeful. They neither possess nor desire to possess worldly wealth.*[1] At the other extreme was the view of Gonzalo Fernández de Oviedo, Spain's official court historian and a sworn enemy of Las Casas, who saw the Indians as *naturally lazy and vicious, melancholic, cowardly, and in general a lying, shiftless people. They are idolatrous, libidinous, and commit sodomy. Their chief desire is to eat, drink, worship heathen idols, and commit bestial obscenities. What could one expect from a people whose skulls are so thick and hard that the Spaniards had to take care in fighting not to strike on the head lest their swords be blunted?*[2]

As the colonial era unfolded, Spaniards embraced Oviedo's view more than that of Las Casas. Antonio de Morga, president of the Audiencia of Quito from 1615 to 1636, stereotyped the Indians as indolent, mendacious, alcoholic wretches. During the 1740s, Juan and Antonio de Ulloa wrote of the natives of the Audiencia of Quito:

> *more defects are to be observed among them than in the other classes of the human species: some are remarkably short, some idiots, dumb and blind, and others deficient in some of their limbs. [Juan Antonio de Ulloa insisted that] Indians are greatly addicted to theft, in which it must be owned they are very artful and dextrous,* [and that they were] *rude, indocile, and living in a barbarism little better than those who have their dwelling among the wastes, precipices, and forests. For if considered as part of the human species, the narrow limits of their understanding seem to clash with the dignity of the soul; and such is their stupidity, that in certain particulars one can scarce forebear entertaining an idea that they are really beasts, and even destitute of that instance we observe in the brute creation. . . . Fear cannot stimulate, respect induce, nor punishment compel them.*[3]

After five centuries of contact, whites and Indians still live apart geographically and culturally. The descendants of the Spaniards continue the culture of individualism and materialism to dominate and control their environment in the name of progress. In contrast, the descendants of the Incas maintain their focus on the communal and spiritual. Humble and group oriented, they seek to live in harmony with nature.

Despite such pervasive and negative European attitudes toward the Indians as those of Oviedo, Morga, and the Ulloa brothers, no other nation developed as humane and as voluminous a code of laws to protect a subject people against oppression as did Spain. No such benevolent body of legislation exists in English or any other colonial jurisprudence of the time as is found in the Spanish Laws of the Indies. While a huge chasm divided theory and practice in the colonies, Spaniards nonetheless struggled among themselves for justice in their conquest and colonization. The king, for example, issued and reissued a decree six times between 1526 and 1548 that stated that the Indians were to be free and not subject to servitude, and the Crown mandated schools in Indian communities, primarily to teach the Spanish language. And while that rule, like so many others, was reiterated in royal decrees throughout the colonial era (it was honored more in the breach than in the observance), the Spanish effort to educate indigenous people came long before the idea of popular education caught on in Europe during the second half of the eighteenth century.

Other factors affected Spanish-Indian relations. Recurrent epidemics of measles and smallpox devastated the native population after the Europeans' arrival. Not only did these catastrophes make Indian labor scarce, but much previously tilled land was left uncultivated as their numbers declined. In control of the lawmaking and judicial apparatus, the Spaniards distributed unused Indian lands among themselves. When the Indian population later rebounded in size after the wave of epidemics, the people found themselves largely landless and forced to seek survival on acreage now owned by Spaniards. They were given small plots called huasipungos to support themselves in return for working the haciendas. A system of debt peonage called concertaje offered the Indians some security and assured the white elite a steady and cheap labor supply.

The Spaniards deemed themselves superior to Indians, blacks, and other people of color in much the same way as they considered their nation superior to all others, their faith the only true religion, and men the natural masters of women. Such were the notions of all Western nations of the era. The indigenous people were seen as children in need of adult Spanish supervision. Several trends and institutions combined to concentrate land and power in the hands of this small white elite, including the decline in the native population and the subjugation of Indian labor through the successive encomienda, tribute, mita, concertaje, and hacienda systems. The huge highland estates were worked by a large Indian underclass. Production met local and regional needs and little else, as trade was limited among the various highland regions as well as between the mountains, the coast, and foreign markets.

The encomienda system gave individual Spaniards the right to collect tribute and labor services from a set number of Indian men. The institution originated in the Caribbean, was carried to Mexico, and was subsequently established in South America. Its purpose was to provide cheap workers and income for white colonists, who in return were to protect the natives and provide them with religious instruction. The arrangement did not involve a land grant, but rather a right to collect tribute in a designated area from adult Indian males between the age of 18 and 50. Inspired both by fear of too strong a nobility in the New World and by humanitarianism toward native peoples, in 1542–43 the Crown tried to destroy encomiendas within a generation by forbidding current holders from bequeathing them to heirs. The effort was met with rebellion throughout the Andes, and the prohibition of inheritance was repealed. There were about 570 encomiendas in the colony of Quito in the mid-sixteenth century, and the majority of Spaniards who held them received roughly 1,000 pesos per year from the system. There were exceptions. In Otavalo in 1549, Sebastián de Benalcázar received an enormous encomienda grant that comprised 1,500 to 2,000 Indians. By 1620 the tribute debts for the encomienda had reached 100,000 pesos. The system was never as strong in Ecuador as it was in other parts of Latin America, and it was rapidly replaced by government-operated collection of tribute from the natives.

Most native workers were originally organized in a forced labor system called the mita, a Quichua word meaning "one's turn." Whereas the Incas limited the term of labor to three months, the Spanish extended it to a full year and broadened its use. The obligation under the Incas in their self-sufficient economy was limited to such jobs as constructing and repairing buildings and roads, whereas the Spaniards sought surpluses for export and used the mita to gain as much profit as possible. The mita continued in textile operations and in agriculture throughout the seventeenth century. Debt peonage (concertaje) gradually replaced the mita and provided a more stable supply of workers through a system of legally enforceable debt obligations in which wage advances were made that most workers were unable to repay. With the bulk of the best land shifted to white hacendados (owners of haciendas), the Spaniards transferred the wealth derived from native labor to themselves. The Indian tribute, moreover, rapidly became the single largest source of government income in the Audiencia of Quito and accounted for 30 to 50 percent of its revenue.

Although Indians were primarily agricultural workers and domestic servants, at the end of the sixteenth century a textile boom began and many were forced to labor in workshops called obrajes. Wool and cotton cloth woven in the obrajes grew into a major export to clothe miners in Peru and Bolivia and made the Audiencia of Quito a dynamic economic center. Coercion of various kinds kept the Indians at their looms, including debt peonage, an obligation rigorously enforced by imprisonment. Weavers often worked from sunup to sundown and sometimes were chained to their looms. By the 1720s the highlands had already entered a period of stagnation when the volume of silver extracted in Peru and Bolivia dropped and European and Chinese textiles began to compete. Textile production diminished from 50 to 75 percent between 1700 and 1800, and the economic decline continued beyond independence from Spain in 1822. Concurrently, the coast began to prosper as cacao became an export of increasing significance.

Church and state were closely intertwined. The Catholic Church held a monopoly on ideology, and more than any other group had the resources to promote education, the arts, and cultural activities. By the mid-eighteenth century the institution was the Audiencia's largest landowner, and the Jesuit Order had amassed so much wealth and power that in 1767 King Charles III expelled it from the colonies. Although there were countless exceptions, many observers claimed that the clergy stressed the outer display and formalities of the faith rather than inner moral development, ethics, and the social consciousness. The Church's material splendor, moreover, stood in stark contrast to the widespread and enormous poverty and exploitation suffered particularly by people of color.

At the end of the colonial era, the coast's advancing prosperity stood in sharp contrast to the relative stagnation and decline that characterized the highlands, where the economy never fully recovered from the textile industry's demise. Spain's Bourbon monarchs took a new approach to promoting colonial economic development, and in 1778 Charles III promulgated a decree of free trade, expanding from six to twenty-four the number of ports in Spanish America permitted to trade directly with each other or with any port in Spain. As a consequence, commerce was thriving in Guayaquil by the end of the century. Most products entered or exited the Audiencia through that port, where shipbuilding began within decades after the Conquest. By the eighteenth century the region's economy expanded when cacao was introduced around the Guayas River basin. The chocolate bean lifted government revenue and eventually made possible a measure of social reform.

In this late colonial period, the opportunities for the highland elite were few. Upward mobility for the so-called criollos, those of Spanish descent who were born in the Americas, was made difficult by the imperial system's preference for Spanish-born bureaucrats and clergymen. Peninsulares, those of Spanish birth who emigrated to the colonies, felt innately superior to criollos and all others, and deemed those born in the New World untrustworthy, poorly trained, badly disciplined, and guilty of vices attributed to the multiracial Americas. The situation worsened in the late 1800s. Whereas in the previous century Hapsburg monarchs sold Crown appointments in the colonies as a means to raise revenue and let criollos buy fifty-one of ninety-three positions as judges by 1750—a full 55 percent of the posts—by 1807 the Bourbon kings, bent on tightening colonial controls, allowed the frustrated peninsular elite only 12 percent of the posts.

The Church was another center of conflict between Spaniards and those born in the Americas. Although criollos constituted a majority of the priests, nevertheless those born in Spain adopted a system that protected them in selecting provincial superiors. The Alternativa, as the name suggests, gave peninsulares disproportional representation, as elections for leaders simply rotated the choices between criollos and peninsulares. Some of the frustrated criollos found prestige and status in the ranks of the swelling militias created in the late eighteenth century to provide internal and external security for the colonies. By 1800 criollo militiamen outnumbered regular Spanish troops severalfold and later served as the military foundation for the independence armies.

When Napoleon invaded Spain in 1808 and gave the occupied country's throne to his brother, Joseph Bonaparte, the issue for many in the Americas was who would rule in the name of the deposed Ferdinand VII, who remained under arrest in France from 1808 to 1814. Quiteños say theirs was the first call for independence in Latin America. On August 10, 1809, a group of criollo estate owners seized power, ousted the peninsulares, and claimed to rule on behalf of the king. Quito's rebels lacked the anticipated popular support for their cause, and as loyalist troops approached, members of the governing junta peacefully handed power back to Crown authorities. On August 2, 1811, another uprising took place, the net result of which was a temporary blending of various political interests in a new governing body. Criollos joined a junta headed by the peninsular president of the Audiencia. A few months later a constitution was issued that vowed allegiance to Ferdinand VII, if and when he should resume the throne with the proviso that the king must respect the institutions and liberties established in the constitution. The junta was soon quashed, however, by royalist forces from Lima.

Spain held firm control of the tense colony until 1822. For centuries Spanish America's political experience had been one of near absolute monarchy, of strong executive authority. Criollos and others did not have any experience with the institution of a national legislature, nor were they accustomed to promulgating laws within the boundaries of constitutional checks and balances. Yet on the eve of Ecuador's independence there was a new tension and division: one between royal absolutism, on the one hand, and the new governing notions as expressed in the constitution of 1812, on the other. Fresh ideas of limited monarchy rapidly evolved into fullblown republicanism, with a three-way division of executive, legislative, and judicial powers. The new notions did not go unchallenged by those more grounded in tradition, and the same strains evident in the 1812 constitution and the king's reaction to it posed a politically destabilizing problem that continued for the next two centuries in Ecuador and elsewhere in Latin America.

The idea of a constitutional monarchy, either in Spain or in the Americas, was swept aside in 1814. The restored monarch declared his authority undivided and annulled the constitution of 1812. His absolutist move and attempt to restore the pre-Napoleonic regime gave impetus to the independence movement in Latin America. Ambitious and stifled criollos now moved away from constitutional monarchy toward the idea of republicanism. The forces that finally liberated the Audiencia of Quito were assembled in the south by José de San Martín, who entered from Peru, and in the north by Simón Bolívar, who swept down from Colombia. Bolívar dispatched General Antonio José de Sucre to liberate Guayaquil, and from there Sucre proceeded to liberate the highlands in 1822. For the following eight years Ecuador was included in Gran Colombia (modern Ecuador, Venezuela, and Colombia). After the Spanish defeat, Sucre was named by Bolívar to head Ecuador, then called the Department of the South, and Venezuelans and Colombians occupied most of its administrative and military posts.

Discontent with the tripartite union of Gran Colombia surfaced from the outset. Ecuador's growing opposition to Gran Colombia was fueled by a 30 percent tax levied on cacao exports to finance Bolívar's final campaign against the Spaniards in Peru and Bolivia, and by a disproportionate 21.5 percent share of independence war debt being thrust upon it. Furthermore, the locals resented outsiders holding most of the top government positions. As the northerners had liberated the region, they determined to dominate it as long as they could.

Bolívar had absorbed the notions of individualism and free-market economics from Adam Smith's *Wealth of Nations* (1776). He hoped that the Indians would be integrated into society by becoming dynamic producers and private businessmen. With that goal in mind he decreed the breakup of communal holdings, which he deemed an archaic form of agricultural tenure and an impediment to Indian assimilation. The lands in theory had been protected by the Crown during the colonial era, and that centuries-old policy was ended by Bolívar's action. In practice, numerous non-Indians manipulated or coerced former communal landowners out of their holdings through such practices as having illiterates sign sale documents in return for a pittance. Informed of both the abuses and the disastrous results, which ran contrary to the desires of the majority of Indians who wanted to maintain their communal life, Bolívar after a few years called his effort a mistake, halted the process of selling off Indian lands, and nullified his own law. He forbade communal holdings from being sold for the next quarter century but did not discard his hope that the indigenous population could be converted to individualism and assimilated.

After Venezuela bolted in January 1830 and the three-part entity of Gran Colombia disintegrated, the independent republic of Ecuador was proclaimed on May 13, 1830. A constitutional convention that named Ecuador's first head of state, General Juan José Flores, wrote a constitution that assured power to a small group of landowners. Flores served three intermittent terms between 1830 and 1845. Tied to the landed elite through marriage, he continued Bolívar's conservative tradition. Voting was restricted to those with property worth at least 300 pesos, and women were excluded from casting ballots and holding office. Congressmen were required to have 4,000 pesos' worth, and the presidential threshold was set at 30,000 pesos. Flores's salary of 12,000 pesos took up about 8 percent of the country's first budget, and the well-paid leader soon owned the country's largest estates.

After independence, regional power was exercised by those who held the great estates. Civil wars were frequent as would-be leaders attempted to create a central state to supersede regional authority. No one could create unity, and property qualifications for voting and holding office assured continuity to a small group who occupied government positions. An enormous social and economic chasm separated the Indian majority from the white elite, and most of the former lived on the margins of economic, cultural, and political power. The conservative Church's monopoly on basic ideology and education confirmed and legitimized the status quo. Owing to the nation's poverty and internal strife, the military took over a huge portion of the budget, consuming 200,000 of the 387,973 pesos expended in 1831, and 530,007 of the 847,657 pesos spent in 1843.

Just as the Andes formed formidable geographic obstacles to political unity, so too did internal barriers hinder communication and trade. The country's three main regions formed largely self-sufficient economies. The coast looked to the world market, the north-central highlands to Colombia, and the southern highlands to Peru. Guayaquil's fierce rivalry with Quito was fueled by resentment that taxes collected on the coast went primarily to the capital to underwrite the central government. Revenue came primarily from cacao and other coastal exports as well as from levies on imports. In subsequent decades, antagonism between the coast and the highlands developed many bases, but the questions of who would pay taxes and how and where revenue would be spent continued to be divisive issues. Another source of regional conflict was competition with the highlands for labor. Workers were scarce on the coast but plentiful in the mountains, where their willingness to migrate was restricted by debt peonage as well as by strong ties to the land. The coastal economy, based on commercial agriculture and aimed toward foreign markets, grew increasingly dynamic compared to that of the country's interior.

A major challenge to Flores came from Vicente Rocafuerte and the so-called liberals in Guayaquil. The solution to the violent conflict between Flores and Rocafuerte was a simple agreement between them to alternate national power. Flores also tried to impose a three-peso tax on all citizens, which was the same amount paid by the Indians in the centuries-old tribute that affected them alone. The measure was resisted nationwide, but particularly in the highlands and among the clergy. A revolt led by the Guayaquil elite broke out in 1845 that drove Flores from the presidency. From Flores's fall until the rise of Gabriel García Moreno in 1861, the country had ten governments and three constitutions. Many of these regimes were led by generals, and those elected continued to be chosen by a wealthy few. Not until García Moreno's 1861 constitution were property requirements for voting eliminated and secret ballot elections instituted.

For decades the largest source of state income remained the Indian tribute. The ethnic-based tax was levied on those least able to pay—the politically excluded sector of society. From the outset of the republic, efforts to tax anyone or anything other than Indians, exports, and imports met stiff opposition. Fierce resistance came from those who deemed themselves superior and apart from the indigenous population, and the Indians alone continued to pay the tax. Its demise came slowly over the course of the nineteenth century and was related to the rise in the coast's economy, whose population also rose in relation to that of the highlands. Cacao was the key. Although introduced to Europe from Mexico in the 1520s, its gradual rise in popularity during the nineteenth century came when Europeans and North Americans learned to sweeten the bitter chocolate with sugar, cinnamon, and vanilla. As a consequence, Ecuador's cacao bean exports soared. Cacao planting gradually expanded north from Guayaquil, and the coastal population grew 2.6 percent per year from 1765 to 1805.

By 1857 customs receipts brought in quadruple the amount raised by the Indian tribute as a result of increases in the price and volume of cacao exports. In the same year the tribute was finally abolished. Cacao revenues also facilitated the gradual abolition of slavery during the first half of the nineteenth century, with compensation to former slave owners. The institution continued until José María Urbina abolished it by decree on July 25, 1851. However, the majority of blacks, despite their new legal status, remained bound to their previous owners through debt tenancy.

Friedrich Hassaurek, the U.S. minister to Ecuador in the 1860s, described rural land and labor conditions which, despite the abolition of the Indian tribute and slavery, had changed little since independence:

> *The Indian farm laborers—and it is only Indians and Negroes who work on farms, and by the sweat of their brows maintain the white population by whom they are oppressed—are called ganaes, or concertados, or peones. Their wages do not exceed a medio (half a real) per day, which would amount to about 23 dollars per year. In addition to this, the landowner is obliged to give to each man a suit of coarse common cloth and a hat every year. He also gives them a small piece of ground which they may cultivate for themselves, and on which they may build their huts, called huasipungos. For this miserable allowance they are compelled to work from early dawn until five or six o'clock in the evening. . . . It is evident that however cheap living may be in the interior, and however limited an Indian's wants may be, half a real per day is insufficient to maintain him and his family . . . the landowner therefore advances the money or furnishes the necessary articles, and he does so willingly, because it is in his interest to keep the Indian in debt. An account is kept of all these transactions. . . . Imprisonment for debt has not yet been abolished in Ecuador.*[4]

Huasipunguero and concertaje workers were required to use their own tools, which because of their poverty differed little from those at the time of the Spanish conquest. The families of Indian workers provided domestic service (huasicamia) to landowners and did so without pay. Landless relatives of huasipungueros who lived on the hacienda were required to work one day per week for the hacendado. If they helped during the harvest, their pay was what they could collect from the scraps left behind in the fields, the yanapas, which gave them the name of yanaperos.

Indians (along with blacks) continued at the bottom of society, a fact reflected not only throughout their lives, but also in death. In 1839, for example, President Flores signed a law that established three classes of racially based payments for Church funeral services. Burial costs for whites were set at twenty pesos; Cholos paid six; and Indians, three pesos. A similar fee structure was set for marriage, and convents often required thirty days of free labor from young women about to marry, a practice called the ponga. Even though Indians and blacks paid far less than those of lighter skin, for financial reasons many Indians did without the wedding ceremony. Some nineteenth-century observers calculated that in part as a consequence of the fees, 75 percent of Indian couples did not formally marry.

Negative attitudes toward the natives existed alongside positive ones, just as they had during the centuries of colonial rule. Hassaurek wrote: *The word Indian is a term of contempt, even among the Indians themselves, who cannot offer a greater insult to one another than by the epithet Indio bruto.*[5] José María Velasco Ibarra, president on five occasions from 1934 to 1972, said that Ecuador's ruling classes *had contempt for the Indian.*[6] Yet neither of these twentieth-century luminaries was himself contemptuous of indigenous people. They agreed with Galo Plaza Lasso, President from 1952 to 1956, who observed that *Indians raise most of the food we eat in Ecuador. What nonsense it is to say that Indians are a drag on the economies of our Andean countries. Forty percent of Ecuadoreans are Indians. Most of the rest are Mestizo. With love and understanding, the Indians are ready to forget four centuries of abuse and increase their stature in our society. We need their common sense.*[7]

The nineteenth century saw indigenous uprisings similar to those of the colonial era against cultural domination, the loss of land, harsh labor conditions, high tribute payments, and the sexual abuse of women. The revolts continued to be isolated and brutally suppressed by white authorities. Fernando Daquilema was one of many Indians who rose in protest against the exploitation of his people. He was captured and shot in 1872 after leading uprisings in Chimborazo province against abuses of mita labor and the tithe.

In the early twentieth century, Eloy Alfaro's liberal government separated Church and state, and the Church was forbidden to collect the 10 percent tithe (diezmo) and a portion of agricultural production (primicia). Its economic power was further reduced in 1908 by the Law of the Dead Hands (Ley de Manos Muertas), by which the Church's vast estates were appropriated and distributed to whites and Cholos. Indigenous communities who were the most in need of land were not among the recipients, but the natives did benefit from other liberal reforms. Although large estates in the highlands were not altered in fundamental ways, the concertaje system of debt peonage was abolished along with imprisonment for debt and a number of land taxes. The huasipungo system survived concertaje's end and served to keep the rural power structure intact. The obligation of Indians to contribute labor for road building and other purposes, called the subsidiary contribution, was halted as well when Alfaro declared that municipal taxes should fund public projects. Many of the liberal leader's followers affectionately called him "El Indio Alfaro."

Currents of change were seen on several fronts as the twentieth century unfolded. Urbanization increased alongside industrialization, thus strengthening the power of urban labor, the middle class, and urban elites relative to the influence of the traditional agrarian power structure. In 1923, for example, the Liberal Party called for an end to the archaic institutions of the huasipungo and concertaje as well as for agrarian reform to eliminate large and inefficient estates. Both systems bonded Indians to large estate owners, which provided the latter with cheap labor. Technological improvements in agriculture, and what was termed rehabilitation of the Indian, were touted as well. Pío Jaramillo Alvarado published *El indio ecuatoriano* in 1922, documenting rural labor exploitation and concluding that Ecuador's most serious problem was the miserable subjugation of the natives. Social justice, he insisted, was a precondition for national economic development. El Inca, one of the first indigenous unions to demand land and labor reform, was organized in Pichincha province in 1927; similar groups soon sprang up throughout the highlands. These and other rural organizations tried to form a united movement in 1931 but were thwarted by the repression of government soldiers and police.

In the countryside and urban areas alike, during the 1920s the popular classes and the left were restless. In September 1923 hundreds of campesinos on the Hacienda Leito in Tungurahua province went on strike for a wage increase. They worked up to twelve hours per day five days per week in return for one real per day and the right to cultivate and raise their own animals on small plots of hacienda land. Most were Cholos whose ancestors had lived under similar conditions since the eighteenth century. They were ordered off the estate and the owner attempted to take their animals. When they refused to leave their homes, local authorities labeled their action a Communist uprising, and President José Luis Tamayo ordered troops from Ambato's military garrison to end the dispute. Hundreds of campesinos were killed in what most historians label one of Ecuador's worst peasant massacres. In the aftermath of numerous urban labor strikes and the Hacienda Leito slaughter in the early 1920s, politicians spoke of the need for social change. Although labor reforms were put on the books, enforcement was minimal.

Prior to 1933, bananas never exceeded 1 percent of the value of exports, but in that fortuitous year United Fruit Company, owned by U.S. citizens, bought an abandoned cacao plantation and began planting. In 1947 the banana bonanza began. As a consequence, between 1942 and 1962 the population of the coast increased by more than 100 percent, whereas that of the country rose by 45 percent. Fueled by bananas, from 1947 to 1957 the Ecuadorean economy grew at an average annual rate of 5.3 percent. After decades of chaos, political stability resulted from the renewed prosperity. But when the surge ended in the late 1950s, due primarily to the recovery of other Central American producers and to stabilized world demand, old patterns of political instability reemerged almost instantly.

As bananas boomed throughout the 1950s, the winners of four presidential elections took power and governed in relative calm from 1948 to 1961. Political parties proliferated and divided the center, left, and right. The popular and middle classes played ever larger roles in the nation's politics, as did manufacturers, industrialists, and bankers, and the United States had enormous influence as the major trading partner, investor, and lender. The Church gained progressive voices, among them that of Leonardo Proaño, who focused on poverty and social justice.

Politics prior to the oil boom of the 1970s was elitist; few Ecuadoreans participated in the election process. Barely 3 percent of the population cast votes in the presidential election of 1888, and in 1925 the percentage was about 11. Only 3 percent took part in the balloting that elevated José María Velasco Ibarra to office in 1933. Until the late 1970s a candidate did not need majority support to be elected but only had to garner more votes than his closest rival. In 1968, for example,

Velasco Ibarra won with 280,000 ballots, less than one-third of those cast in a field of five. Politics was highly fragmented and for the most part centered on individuals rather than on parties or principles.

Velasco Ibarra was the central figure in Ecuadorean politics from his first term as head of state in 1934 until 1972, his last year in power. Almost invariably the fiery orator launched ambitious public works programs, most commonly roads, bridges, schools, and electric plants, and increased spending on education. The bureaucracy swelled during his administrations and military outlays rose. He usually found himself trapped between growing demands for expenditures, on the one hand, and limited finances, on the other, and would assume dictatorial powers in the face of congressional or popular opposition. Velasco Ibarra's five elections remain a Latin American (and perhaps a world) record.

The Ecuadorean military, though active in influencing politics throughout the century, made two major interventions prior to the 1970s. On both occasions those in uniform provided reform regimes. The first came in 1925 with banking and tax restructuring. The second occurred in 1963, with an administration that inaugurated a modest land reform, added a personal income tax to the corporate levy, attempted to centralize revenue collection, and raised import duties to reduce the deficit. It was not the last time that soldiers would influence politicians.

3

The Oil Era

Ecuadoreans have known the utility of oil for centuries. Before the Spaniards arrived, the sticky substance was turned to tar by coastal Indians who used it to caulk canoes, waterproof arms and utensils, and make torches. In their simple approach they dug pits that filled with oil, and in time the pools evaporated into tar. Most of it came from the Santa Elena peninsula near Guayaquil, and the Spaniards used it for tarring their own ships by adopting the methods of the natives. Modern wells began operation on the peninsula shortly after World War I. Although their output was always low, a more promising search was launched in the Amazon region in 1921 when Standard Oil obtained the first concession to explore there. However, decades passed with few discoveries and little production, leading President Galo Plaza Lasso to dampen expectations that Ecuador would someday join the ranks of oil-producing nations. *The Oriente is a myth,* he concluded in 1949; there simply was no oil to be found in the Amazon.[1]

The former president was proven wrong in 1967 when a Texaco Gulf consortium discovered vast amounts of crude in the far north of Sucumbíos province in the Oriente. Oil launched Ecuador's third economic boom of the century, following those of cacao and bananas, when on June 26, 1972, it began flowing through the 312-mile-long Trans-Ecuadorean Pipe Line (SOTE) that rose 13,000 feet from Nueva Loja in Sucumbíos, went over the Andes, and then descended to the Pacific port of Balao in Esmeraldas province. From there it was shipped to foreign markets, primarily that of the United States. The pipeline was privately built and was to be privately operated for twenty-five years before being turned over to the state. The line initially handled 250,000 barrels per day, but by century's end it was enlarged three times to keep up with production and demand. In 2000 it carried 360,000 barrels per day. Although for the next three decades Ecuador would never produce over 0.5 percent of the world's crude oil, in 2001 the government relied on the revenue it produced for about 46 percent of its income.

PROTECTING A FORCE FOR CHANGE

In 1972 a section opened of the 312-mile-long oil pipeline that rises from the Amazon to cross the Andes at 13,000 feet before descending to the northern Pacific port of Esmeraldes. Oil's transformation of Ecuador came with a price. In March 1987 an earthquake swept away twenty-five miles of the pipeline and an environmental disaster, worse than the *Exxon Valdez* catastrophe, followed.

In the late 1950s worldwide petroleum production far exceeded demand. Private companies controlled its price until the 1970s, when the Organization of Petroleum Exporting Countries (OPEC) succeeded in manipulating the supply. OPEC was launched in 1960 with headquarters in Vienna for the purpose of expanding revenue in oil-producing nations. Ecuador joined its ranks in 1973. In the early 1970s the cartel reduced oil exports to non-OPEC countries, thus creating an international shortage and higher prices in the new seller's market. Navy captain Jarrín Ampudía, Ecuador's minister of natural resources, was elected OPEC's president in 1974 when it held its annual meeting in Quito. For a time the country complied with production ceilings on how much crude could be produced by each member.

The oil bonanza lasted a decade and a half as prices skyrocketed. In 1972 the cost per barrel for Ecuadorean crude stood at $2.50, and the next year it came close to doubling at $4.20. By 1974 the charge soared severalfold to $13.70. It climbed to $23.50 in 1979 and reached its summit in 1980 at $35.26 following the Iranian revolution and conflict in the Middle East. Booming prices then tapered off a bit but still remained high and ranged from $34.48 in 1981 to $25.90 in 1985. In 1979 oil export earnings catapulted 85 percent over the previous year. Black gold transformed Ecuador as rapidly and profoundly as anything in its history except the Spanish conquest.

The petroleum surge, unlike the earlier cacao and banana booms, occurred beyond the coast. It came from the Oriente, and the revenue it generated went directly into government coffers in Quito. The newfound wealth financed unprecedented spending and in fewer than thirty years changed the face of the country. In the decade following 1972, public expenditures accelerated at the astonishing rate of 12 percent annually and rose from 22 to 33 percent of the gross national product. Oil lifted foreign exchange earnings eightfold between 1971 and 1974 and almost overnight provided nearly one-half of all government revenue.

The rapid development paralleled the cacao and banana eras in that the state continued its dangerously high dependence on a single export as the source of most of its revenue. That reliance kept taxes relatively low for most citizens, and during the 1970s and early 1980s the relative burden of taxes fell. The principal instruments for income redistribution—income and property levies—declined as the number of citizens paying them dropped by about 6,600 from 1972 to 1976. Foreign oil purchases helped keep domestic taxes at low levels, and Ecuadoreans grew accustomed to prosperity and development. Domestic petroleum products were heavily subsidized, and their internal consumption rose dramatically along with the number of automobiles on the nation's increasingly congested streets. Most Ecuadoreans cooked with state-subsidized tanks of natural gas, which in 2000 cost about $5 to produce and sold for $1. Oil dependency, like the earlier reliance on cacao and bananas, reinforced established attitudes toward taxes and spending that proved hard to break when, at least in relation to need, state petroleum income declined in later years.

President José María Velasco Ibarra, who worked without a congressional majority after his 1968 election, broke the country's political impasse by assuming dictatorial powers in 1970. He proceeded to raise taxes, close universities that had become centers of opposition, and increasingly lost public support. On February 15, 1972, General Guillermo Rodríguez Lara headed a coup d'état that ended the president's fifth and final administration. Army officers determined that corrupt politicians would not make good guardians of the nation's recently discovered wealth and believed that the military was entitled to share in the abundance. The generals spoke in nationalistic terms of Ecuador's unlimited future and raised hopes that through oil the country would join the ranks of modern nations. The military regime's first major action to meet heightened expectations was to increase the state's share of oil revenue, and as prices rose sharply from 1972 to 1980, so also did state income, which was now many times greater than during previous presidents' terms. It was five times more then Velasco Ibarra's regime took in, and twenty times as much as Plaza Lasso's. It shot up fourfold in the first three years alone. In the 1970s, Ecuador's economic output rose from $4,347 million to $10,155 million, a stupendous 233 percent increase when compared in constant 1990 dollars adjusted for inflation. The rate of economic growth was double that of the previous decade. Local governments, rural electrification programs, universities, and a host of other entities received a portion of oil revenue, but the military took far more than the others. From 1972 through 2000, 45 percent of this revenue went to the Armed Forces. The money was not included in the country's legislatively approved budgets and was removed from congressional oversight. The Army created its own industrial enterprise (DINE), the Navy launched a petroleum fleet (FLOPEC), and the Air Force ran its own airline (TAME).

During the 1970s and 1980s, Ecuador gradually took over the Texaco Gulf consortium, followed by state ownership of other enterprises. The first step came in 1971 when Velasco Ibarra formed the Ecuadorean State Petroleum Corporation (CEPE). Within a few years it acquired a 25 percent interest in Texaco-Gulf, and later the level reached 51 percent. In 1976, CEPE took over the installations of a British company, Anglo Ecuadorean Oil Fields Ltd. State expansion continued and broadened as an interest was acquired in the oil refinery in Esmeraldas that went into operation in 1977, and the following year FLOPEC was launched to transport petroleum. President Rodrigo Borja finally took over the last shares of Texaco-Gulf in 1989, and shortly afterward CEPE was restructured and renamed Petroecuador. He denied that the buyout of foreign oil facilities amounted to nationalization, despite company protests to the contrary, and insisted that foreign investment and expertise remained essential for modernization and expansion.

For the first time in Ecuador's history, with significant state ownership in the oil industry and the ability to tax private foreign petroleum companies, governments had their own source of revenue. The financial independence diminished the power of the traditional oligarchy, who had controlled much of the country's economic and political resources. The military regime not only had the state run an oil company, refinery, and fleet but also built roads at an unprecedented rate, provided credit for agricultural development, supplied incentives for manufacturing and industry, and acquired sophisticated equipment for the Armed Forces. Economic expansion in Quito stimulated the growth of the urban middle class and in the process attracted people from the provinces. Ecuador's predominately agricultural economy was transformed into one in which services, manufacturing, industry, and mining played ever-expanding roles. The bureaucracy burgeoned the most, and thousands of public service jobs were created. Whereas the population grew from 6,521,710 in 1974 to 9,648,189 in 1990, an increase of 27 percent, the number of government employees swelled fivefold from 100,000 to around 500,000 during the same decades. In the process, the middle class and business interests gained greater political assertiveness.

Those who directed Ecuador saw oil as an opportunity to modernize, a means to escape both underdevelopment and poverty, and a way to build a dynamic, developed, and industrialized economy. In the 1970s they embraced import-substitution measures to stimulate manufacturing and industry, increase domestic employment in the growing cities, and prevent social unrest. The policies were an alternative to the export-led strategy for economic growth, and the approach had several facets. High import taxes were imposed to protect domestic businesses against stiff foreign competition and to discourage cheap imports. A portion of oil revenue was invested to launch state-owned industries or to purchase shares in existing enterprises deemed essential to development. Before 1972, state involvement was limited to the production of fertilizer and a few other industries, and small state enterprises accounted for less than 2 percent of the gross national product. In contrast, by 1983 the contribution of public undertakings had shot up to 12 percent, with most government involvement being in the areas of petroleum, transportation, communications, and utilities. The regime also employed heavy state subsidies to stimulate consumer demand and encourage economic expansion. By 1978 almost one-half of the budget was devoted to food and energy subsidies, exemptions, and credits.

The generals devised elaborate plans built on earlier ones to revamp and modernize the republic. During the 1950s the state tried to foment industrial development; the 1957 Industrial Development Law was one such attempt, and similar ones were made by the military junta in 1963. The exertions had been relatively limited for lack of financial resources. In 1972 the Junta of Economic Planning and Coordination presented what it called the Fundamental Outline of the Integral Plan for Transformation and Development, followed in 1973 by the Five-Year Development Plan. The latter urged the direct participation of the state in the productive process through the introduction of basic industries, which implied that the public sector should adopt a very active stance in the promotion of enterprises. The nation's executive, General Guillermo Rodríguez Lara (1972–1976) further explained the government's objectives. Ecuador, he said, required decisive intervention by the state in the economy to promote reforms necessary for national development. The strategy, he argued, meant transferring from foreign hands to the public sector the fundamental decisions affecting the economy and society.

Modernization attempts at first seemed to work well. Industry expanded rapidly. During the 1970s it surged ahead at an average annual rate of 10.5 percent and by 1980 accounted for 19 percent of the gross national product. Public enterprises played a significant part of the expansion. Whereas in 1972 the value added to the gross national product by state undertakings was 2 percent, by 1983, as noted, it had catapulted to 12 percent. New factories mushroomed around Quito and Guayaquil, with chemical, wood, paper, and metal-processing plants among those that led the growth. The Ecuadorean Development Bank was created in 1979 to provide credit as high levels of domestic investment were needed for machinery and parts to advance industrial production. The state provided cheap domestic gasoline, cooking fuel, electricity, and bus fares. The enormous expansion in road construction resulted in increased commercial exchange and movements of people between regions. One striking example was the blossoming of the flower industry in the highlands during the 1980s due to better roads and cheap transportation costs.

A new myth emerged, that of the virtually never-ending oil fields of the Oriente. Some said that the reserves were greater than Venezuela's and were comparable to those of the Middle East; the end of the boom was simply nowhere in sight. Ecuadoreans used a substantial amount of their own petroleum directly: the number of automobiles rose from 82,000 in 1970 to 223,000 in 1977, in part because state-subsidized gasoline was sold to the public at less than one-half of the cost of production. Hefty government subsidies were not restricted to fuel consumption, moreover, but were extended to include an array of what came to be seen as essentials. Within little time the underwriting imposed a significant burden on the budget.

In 1974, General Rodríguez Lara began to lose support from an impatient public who believed that the oil wealth was not trickling down fast enough. Rural and urban unions felt stifled in their demands, the business community resented a loss of power, and political parties clamored for a return to democracy and an end to heavyhanded governing methods. General Raúl González Alvear led an aborted uprising against Rodríguez Lara on September 1, 1975. Thirty soldiers were killed, and Alvear sought asylum in the Chilean embassy. Five months later on January 11, 1976, Rodríguez Lara was eased out by the heads of all three branches of the military. He resigned and was replaced by a Supreme Council composed of Vice Admiral Alfredo Poveda Burbano as president, Army General Guillermo Durán Arcentales, and Air Force General Luis Leoro Franco. Discontent and repression continued despite Rodríguez Lara's ouster. In 1976, Monsignor Leonidas Proaño, an outspoken advocate of reform and Indian rights, was arrested for conducting a conference of priests in Riobamba to analyze Latin American conditions. That same year a campesino leader was assassinated in Chota in Imbabura province.

Demands for a return to democracy swelled. After seven years of military rule, power was transferred to civilians. A new constitution was written and elections were called. The 1979 charter gave the vote to illiterates, established a runoff round of elections between the top two presidential candidates in the event no one garnered 50 percent of the vote, and created a unicameral legislature. Civilian Jaime Roldós was elected president in 1979. At Roldós inauguration, outgoing head of state Poveda Burbano pointed proudly to impressive indicators of seven years of economic growth during the oil bonanza and military rule. He noted that the budget had expanded 540 percent and that exports as well as per capita income had increased 500 percent. Industrial development, he said with pride, had been impressive, and Ecuador had progressed dramatically in less than a decade. The admiral did not refer to any signs of significant problems. He did not mention that when oil revenue and Ecuador's own capital proved insufficient to sustain ambitious import-substitution and advance modernization plans, the military regimes had responded by borrowing from abroad and using oil reserves as loan guarantees. The foreign debt had swelled twentyfold during the 1970s, from $209 million in 1970 to $4,167 million in 1980. Whereas the government balanced its books in 1970, thereafter expenditures virtually ceased to be on a par with income. In 1971, $15 million more was spent than was taken in, and by 1980 expenditures surpassed revenues by $642 million.

As the 1980s unfolded, the economy, the state, and taxpayers alike continued to rely on oil and international lending agencies to pump up revenue and sustain growth. The bottom fell out of oil prices. Their descent was as dramatic as their rise, and along with the plunge came major changes in public policy. After the oil boom began, Ecuador took three approaches to the economy until the end of the century. The first was from 1972 to 1979, when the state continued to oversee its direction and development. The second, a reformist middle course, was pursued between 1979 and 1984. Finally, the movement of all governments from 1984 to 2000 and beyond has been in the direction of ever-greater free market policies, a strategy that many have dubbed neoliberalism.

In the last fifteen years of the twentieth century, Ecuador suffered through a sharp drop in the price of crude oil, underwent several wrenching recessions with soaring inflation and major increases in interest rates, and saw its foreign debt rise. An earthquake caused an environmental catastrophe by severing the main oil pipeline for five months, and the weather turned unruly as well when El Niño's wild winds and rains caused widespread crop, housing, and road damage. The process of globalization had its impact on Ecuador just as it did everywhere else. These factors created the backdrop for the country's political turmoil at century's end.

Problems in the mid-1980s took away the glitter and gloss of the previous decade's economic expansion. Oil prices dropped, slightly at first and then seriously in 1986. Ecuadorean crude sold for $34.48 per barrel in 1981, declined to $25.90 in 1985, and plunged to $12.70 in 1986. Thereafter prices fluctuated around $15 until 1998, when they fell to $9.15 per barrel. The cost of oil rose in 1999 to a little over $20 per barrel, and the roller-coaster ride continued to 2000. Ecuador benefited from OPEC's decision, made after the 1998 price plummet, to reduce output by 5.2 million barrels per day, and by February 2000 the cumulative impact of the cutback was to boost its crude to $25.80 per barrel. By March 2000 oil prices had virtually tripled in fourteen months to reach a nine-year high. The world consumed about 77 million barrels of oil per day, but production fell to 75 million barrels due to a 4.3 million-barrel-per-day cutback by OPEC. The average world price was $10.72 per barrel in December 1998 and $31.77 in March 2000. For every $1.00 per barrel increase, the state gained an additional $80 million in annual revenues. The Esmeraldas refinery processed up to 110,000 barrels per day in 1999, and its expansion aimed to double that capacity within a year. A measure of disillusionment crept into Ecuadorean thinking insofar as most exploratory drilling for new oil fields was unsuccessful. Some analysts projected that the country's oil reserves would not last past 2010. Nobody knew for sure, and only costly exploration would tell. Nonetheless, in 2000, Ecuador was still Latin America's second largest oil exporter after Venezuela, and its reserves of natural gas, though undetermined, appeared to be considerable.

Throughout the 1980s and 1990s, the country responded to dwindling oil prices by increasing production, hoping that selling more for less would maintain government revenues. New oil fields were sought and found, and output was upped one-third between 1987 and 1991. Ecuador's rise in volume led to conflict with OPEC, which tried to stabilize prices by limiting output. After repeatedly exceeding its quota, Ecuador finally left the oil producers' organization in 1992. It intended to further boost production—to double it to 575,000 barrels per day by 1996—but because of limited pipeline capacity the level reached was closer to 360,000 barrels. Nonetheless, oil exports and domestic fuel sales were expected to finance 45 percent of the state budget in 2001. Petroecuador accounted for about 90 percent of oil exports. It was confident that it could increase its output, but with only one pipeline through which state and private production was channeled, transport capacity was limited. Throughout the 1990s presidents had sought to built another line, but political divisions thwarted the plan. Not until December 2000 was a contract for a second pipeline approved. It was scheduled for completion in 2002.

Petroleum's downward plunge had a huge impact on development policies. Import substitution had fostered a large measure of modernization, but it had not created the dynamic industrial sector that some observers had predicted. In the 1980s, the remarkable industrial growth of the 1970s came to a virtual halt as the sector's output advanced a mere 0.2 percent per year. Little foreign investment entered the economy as the market was deemed too small for significant profits. Moreover, investors considered Ecuador's industry overprotected and noncompetitive compared to that of neighboring countries. Still, import substitution continued to have its supporters, who claimed that high-tariff walls and state involvement in basic enterprises were both essential to development, and it also gave governments greater ability to influence economic and social policy toward agreed-upon objectives.

Throughout the 1970s the state used a portion of expanding oil revenues to amplify social services and subsidize a variety of necessities. A measure of wealth was thereby redistributed and the buying power of most people, for basic purchases at least, was kept relatively stable as the state underwrote the low prices of many food staples, natural gas for home cooking, gasoline for vehicles, electricity, transportation, and other items viewed as essential. As the 1980s ended and the 1990s unfolded, political problems multiplied when oil prices continued to unravel. The increasingly well-organized public fought to maintain subsidies while consecutive administrations struggled to reduce state underwriting that they thought the nation could ill afford.

With oil prices down, the economy slowed, and state revenues no longer bounding ahead, Ecuador entered a vicious cycle of inflation and recession characterized by domestic deficits and a growing foreign debt. Growth after 1985 was weak in comparison to that of the previous fifteen years. Output in 1970 stood at $4.3 billion; and fifteen years later, in 1985, it towered at $11.0 billion. It had soared 225 percent. In marked contrast, the value of goods and services in 1986 was $11.3 billion and reached $14.5 billion in 2000. (The figures are adjusted for inflation using 1990 U.S. dollars.) The increase was 127 percent, a big drop from the preceding surge of 225 percent over the same number of years.

Government expenditures for the remainder of the century consistently outweighed revenues. In 1985, Ecuador had its last budget surplus—$76 million—and every year thereafter the deficit ran several times that amount. In 1970, just before the oil boom began, the foreign debt stood at $209 million and was considered manageable. Although the civilian-led governments that replaced the military after 1979 slowed its rate of growth, the obligation grew over threefold from 1979 to 1989, rising from $3.5 billion to $11.5 billion. From 1987 until 2000, year after year, the debt was virtually equal to the gross national product. In 1990, after the extraction of 1.5 billion barrels of oil from the Oriente, which some claimed was one-half of the country's total reserves, Ecuador had a foreign debt of more than $12 billion. By 2000 the obligation stood at $13.7 billion in an economy with a gross national product of $14.5 billion.

Readily available foreign credit induced governments to accept large high-interest loans during the decade of the 1970s. Ironically, high interest rates, which reflected double-digit inflation, were fueled in large part by OPEC's price hikes. Lenders and borrowers alike were confident that the country could repay the advances with ever-expanding oil revenue. They were wrong. When petroleum prices plummeted, Ecuador failed to reduce spending and restrict state involvement in the economy. The decade of the 1980s saw stagnation under the burdens of debt, inflation, and volatile oil prices.

In seeking new loans or renegotiating old ones, every administration from 1981 forward enacted an austerity program of one type or another. The belt-tightening was frequently mandated by the International Monetary Fund (IMF), the World Bank, the Inter-American Development Bank, the Andean Development Corporation, the Paris Club, or some other international lender, and their urging included selling off state enterprises judged too costly and inefficient, reducing government subsidies, and bringing outlays more in line with income through a combination of raising taxes and slashing expenditures. During the 1980s the IMF was asked for balance-of-payments support on a number of occasions. And while assistance was usually forthcoming, it carried conditions that often included unpopular restrictions in federal expenditures. The government cut education and health services, reduced subsidies on consumer goods including gas, electricity, and basic food items, and eliminated thousands of state jobs.

The business community labeled as dangerous leftists the country's two presidents from the early 1980s. The opponents of Jaime Roldós (1979–1981) accused him of draconian measures when he elevated the price of gasoline threefold in January 1981, an action that caused transportation costs and then those of basic food prices to rise. He adopted the measure after a costly five-day confrontation with Peru that same month, which ended with nearly 200 dead and a hefty increase in the deficit and foreign debt. Doubling the minimum wage, reducing the work week from forty-four to forty hours, and allowing retirement for women after twenty-five years on the job were among his major social reforms. Roldós's successor, Osvaldo Hurtado (1981–1984), of the Popular Democracy Party (DP), governed when an economic crisis swept Latin America. He signed an accord in March 1983 for $166 million with the IMF, which insisted on deficit reduction from 5 to 4 percent of the gross domestic product. Oil prices fell, and in 1982 revenue from that source declined about 11 percent. Hurtado then attempted to cover the shortfall by rigorous tax enforcement and higher rates on the rich. In 1981 over 700 businesses were charged with tax fraud. Interest rates rose, the sucre was devalued by one-third for the first time in a decade (the president devalued the currency on three occasions), restrictions were placed on imports, and the price of gasoline and other necessities increased. A series of five nationwide strikes swept the nation in addition to frequent work stoppages by educators, health-care employees, and transport workers. Inflation rose 53 percent in 1983, when continuous rains caused by El Niño devastated agriculture and much of the nation's infrastructure. Hurtado's social program included worker representation on management boards, greater job security, and a labor code for rural workers.

Hurtado's harshest critics were on the right and a business-oriented chief of state followed, one committed to reducing the role of the state in the economy, ending controls on exports, imports, and interest rates, welcoming foreign investment in all areas including oil, and eliminating all but a few subsidies. The term of León Febres Cordero's (1984–1988), a member of the Social Christian Party (PSC), coincided with a worldwide conservative trend as Ronald Reagan was elected in the United States and Margaret Thatcher in England. Like them, he believed that if business did well, the benefits would trickle down to the rest of society. Febres Cordero had directed the Noboa group of companies, the nation's largest business, and had headed the Guayaquil Chamber of Industry. He obtained $105 million from the IMF in 1985 on the condition that he adopt a restrictive monetary policy. The next year he asked for an additional $91 million but received only one-fifth of his request because fiscal policies fell short of the lender's demands, even though the president upped gasoline prices 80 percent, transportation fares 40 percent, and, in compliance with IMF requirements, set the minimum wage at 15 percent less than Congress approved and 43 percent less than labor demanded. Febres Cordero put through his unpopular austerity measures by flooding Congress with detailed emergency measures that became law if not acted on within two weeks. Most were barely debated. During his term, petroleum prices dropped drastically from $27 to $8 per barrel, thus forcing the government to suspend debt repayments. The vice president called the consequences the worst financial and economic crisis in Ecuador's history. The March 1987 earthquake caused the nation's only oil pipeline to burst and halted exports for five months, during which time no revenues were received from the state's main source of money. During Febres Cordero's administration, the rate of inflation advanced, the deficit increased, and unemployment reached new heights. His major social legislation was to expand health care for mothers and their infants.

Belt-tightening continued throughout the 1990s. Rodrigo Borja (1988–1992), of the Democratic Left Party (ID), criticized his predecessor's neoliberal policies as socially destructive and argued that a good measure of state control was essential to prevent private sector excess and to ensure a better distribution of wealth. Increasing purchasing power, he believed, was the best way to stimulate businesses and advance the economy, yet he continued, albeit more gradually, the policies of raising rates for basic services, lowering subsidies, and devaluating the currency. Borja obtained $135 million in August 1989 from the IMF and another $136 million in February 1991, but because his policies were deemed insufficiently austere, only 86 percent of the first loan and 25 percent of the second were forthcoming. Petroleum prices stayed low, inflation continued its upward advance, and labor strikes remained common. Borja's major social achievement was a massive literacy campaign, but he was unable to fulfill the pledge of his party to pay the social debt by improving health and other social assistance programs.

A self-described conservative followed Borja. Sixto Durán Ballén (1992–1996), of the United Republican Party (PUR), promised to reduce the deficit to 0.5 percent of the gross domestic product and was advanced $180 million by the IMF in March 1994. The money was forthcoming, but the country failed to meet the lender's conditions. Congress blocked the president's IMF-backed proposal to raise the value-added tax from 10 to 18 percent and expand it to include street vendors. Still, telephone and electricity rates were upped 50 percent and fuel prices hiked 115 percent to meet budget shortfalls. A series of strikes followed the price increases. Durán Ballén had counted on oil at $17 per barrel to finance half the national budget, but the price fell to $10 due to a worldwide petroleum glut.

Abdalá Bucaram (1996–97), of the Ecuadorean Roldista Party (PRE), claimed he was a populist and appealed to those on the margins of society, but he pursued belt-tightening as he sought to obtain additional foreign loans. Jamil Mahuad (1998–2000), of the Popular Democracy Party (DP), did the same. Some analysts claimed that the names of presidents and their political parties differed much more than their economic policies.

Social problems multiplied as services waned, prices rose, and the state's financial ability to confront them diminished. The austerity measures were correlated by many critics to rising unemployment and underemployment. The result was social unrest, increasingly well-organized protests and demonstrations, and countless strikes. Government figures set Ecuador's rate of poverty at 47 percent of the population in 1975, 57 percent in 1987, 65 percent in 1992, and 67 percent in 1995. The World Bank, using slightly different methods of calculation, put the figure at 60 percent in the early 1990s and classified two million of six million poor Ecuadoreans as indigent. At the same time that poverty increased, wealth became more concentrated. In 1988 the World Bank concluded that the wealthiest 10 percent of the population had 47 percent of the income. In 1993 they had 54.7 percent. On the other hand, the bank claimed that the bottom 20 percent received 2.55 percent of the income in 1988, a figure that shrank to 1.68 percent in 1993.

Ecuador's debt problems were similar to those of many Latin American nations. During the 1970s, high-interest loans hiked the region's external debt from $30 billion to $240 billion, and in the 1980s it soared to $431 billion. During both decades, debts were renegotiated and restructured, and borrowing and lending continued. But increasingly as the 1980s unfolded and continuing through the end of the century, austerity measures were imposed to cut inflation and deficit spending, and economic growth slowed to a near standstill. The social impact was devastating. During the decade of the 1980s alone, there was a 27 percent increase in poverty.

In 1994, Luis Macas, president of the Confederation of Indigenous Nationalities of Ecuador (CONAIE), which brought together Indian organizations in the coast, highlands, and Oriente, gave his critique of government austerity measures and international lending agencies with their impact on indigenous people and others:

> *We are living in a process of structural adjustment in which the rise in prices for necessities affects all Ecuadoreans, but the situation is even more serious for indigenous peoples who do not have any insurance, salary, or other protection. I think that the government follows the directives of the World Bank and the IMF very closely, and these are policies that impact indigenous peoples throughout Latin America. The Ecuadorean government has to accept the conditions of the IMF and the World Bank in order to obtain new credit. And it does not matter if this negatively affects a great majority of Ecuadoreans. What matters is that they do what is necessary to obtain the credit. These are policies imposed from outside, but they create problems inside our country. It is really part of a global problem that is very complicated. But we are questioning the priorities of multilateral banks and government.*[2]

The country's economic problems and the frustration that they caused continued unabated during these years. By 2000, expectations were largely dashed that the economic and social changes generated during the oil boom would transform the political culture. Traditional aspects of civilian politics continued apace, including those of regionalism and of multiple parties centered around an individual. The number of social, economic, and political groupings proliferated, and intense rivalries between them as well as conflicts between presidents and Congress persisted in fragmenting and weakening the political process.

As bad as the economy was, it got worse. Politics followed suit. National strikes were commonplace. During the 1980s the leading labor union, the United Workers Front (FUT), established in 1971 to unite the country's three main labor confederations and a number of minor unions, led the shutdowns. There were far more strikes during the 1980s than there were during the previous decade. The oil-generated prosperity of the 1970s made it easier for the state to meet conflicting social demands, but from the mid-1980s onward the same task was virtually impossible. In 1990, CONAIE emerged alongside the FUT as one of the major movers in the protests. The FUT maintained a working-class orientation with its focus on improving wages and gave relatively little attention to the claims of other sectors of society. In contrast, CONAIE reached out to a variety of groups and organizations in an effort to advance interests other than its own.

The decline in oil prices alone did not account for the shift in Ecuador's public policies from the mid-1980s onward. Of great significance was the worldwide transformation that restructured almost every economy. Technological change and economic integration, while hardly new forces, suddenly surged to unite the world with unprecedented rapidity. The process went by many names, including "globalization," "the new world economy," and "the information revolution." Whatever one calls it, Ecuadoreans, like everybody else, were far from immune to what it wrought. The process of globalization shifted into high gear in the 1980s, driven by technology—above all by the spread of computers and by rapid advances in telecommunications. It catapulted during the 1990s as the Internet and other technologies created radically new efficiencies and economies of scale. The barriers separating markets and countries came tumbling down, as did the Berlin Wall erected during the Cold War. As they fell, spiraling progress in technology, information, and new forms of finance came together to make knowledge easily available about countless products and services as well as a myriad of opportunities crucial for Ecuador and other so-called developing nations for investing, borrowing, and lending.

Competition increased with globalization and the information revolution. With a personal computer, a phone line, and worldwide delivery systems, new businesses could be started almost instantly, while old ones could revamp and streamline, often at little cost. Buyers now could do comparative shopping all over the globe, and do so rapidly, easily, and inexpensively in cyberspace. The information revolution heightened expectations for better standards of living, as countless people were increasingly able to compare products and services available to them in their own societies with those enjoyed virtually anywhere else in the world. But the accelerated competitiveness devastated many businesses and countries, particularly those that were undercapitalized or insulated from competition by government protection. Both conditions characterized most of Ecuador's state-supported enterprises, which had enormous difficulty in increasing the quality of their goods and services, productivity, wages, and competitiveness. Although not the only cause, among the major reasons for their poor performance was a lack of adequate capital—a deficiency that worsened as oil prices plummeted during the 1980s and continued to plunge into the next decade.

During the 1980s, most developing countries lifted capital controls and opened their markets to foreigners. Instead of a few banks making international loans, many bankers, and then mutual and pension funds and individual investors, became major money suppliers. Vast new sources of capital suddenly became available. All around the globe, countless investors shifted their money with astonishing rapidity, and managers of pension and mutual funds looked for the best investments they could find throughout the world. Some of the latter created separate "emerging market" funds as well as ones targeted at telecommunications, electricity, gas, and other sectors. Enormous amounts of capital became available to the most efficient and profitable.

Backward, bloated, and profitless enterprises, however, were starved for capital to revamp and modernize—unless they had government support for increasingly costly investments to keep pace with the best the marketplace offered. The new money followed few if any political imperatives, as it came not from governments or big banks under state influence but from countless individual and group investors who were simply seeking profits. And if the mass investors largely ignored some countries, if its enterprises offered meager profits, if a government could not control inflation or deficit spending, then it had to look to the IMF and to other international lenders. Loans came with a high price. Lenders imposed rigorous fiscal rules and demanded free-market reforms, hoping that with their acceptance borrowers would finally attract investors. If a regime did not make the necessary adjustments to the new world economy, then it likely fell behind. There was no Soviet Union or socialist bloc to extend a subsidizing hand.

In 1989, moreover, the debts of various Latin American countries held by major commercial banks were converted into U.S. government-backed Brady bonds, many of which were then sold to the general public through mutual funds. Countries with the best economic performances were the ones that attracted bond purchasers, and banks received U.S. government guarantees to extend new loans on the condition that the receiving countries make major economic reforms. That contingency invariably meant moving more toward the free market and in the direction of so-called neoliberalism.

A large portion of Ecuador's economic activity was heavily subsidized, costly, and inefficient, owing in large part to heavy state involvement in the economy through the industrial import-substitution policies of the 1960s and 1970s. Much of it was perceived as far from adequate in providing the kinds of goods and services of acceptable quality and price that an ever more knowledgeable public demanded. Ecuador, along with the rest of Latin America, adjusted to globalization in ways similar to that of most of the world, albeit at its own pace. In the late 1980s and increasingly during the 1990s, its public policies shifted in free-market directions, toward the reduction or elimination of subsidies and the privatization of enterprises owned and operated by the state.

In Ecuador, as everywhere, the transition from the old to the new economic order was far from smooth. Many people resisted the destruction of old enterprises, jobs, and ways of living and of looking at themselves and the world. Still others feared globalization's disruptive impact on culture, tradition, society, and politics, and on who and what they were as individuals, groups, and nations. By the end of the century, the central movement was clearly in the direction of free-market reform, government downsizing, cost cutting, reducing inflation, and attracting foreign investment. Yet in 2000, Ecuador had still not formed a solid consensus on just how and at what pace to adjust to the new rules of globalization. It is within that huge, multifaceted global context that the country's politics, economics, and society of the last two decades of the twentieth century must be viewed. Likely the same lens will be useful well into the new millennium.

In August 2000, President Gustavo Noboa looked back on the last decade. He summed part of it up by saying that in 1992 his predecessor, Durán Ballén, had begun the process that led to Ecuador's participation in a globalized, free-market economy. Since that time, he said, *the necessary structural reforms were not realized to deal with the new reality. The failure to adopt those changes has led Ecuador to a serious solvency problem.*[3] In campaigning for and gaining the nation's highest office, Jamil Mahuad in 1998 was more forthright about his intentions to impose austerity measures than was Abdalá Bucaram when he won the presidency a few years earlier. While their election approaches diverged, the central fact remains that they offered only marginal differences in their approaches to strengthening the free market and imposing austerity measures. They simply saw no alternative. The only other option was to stagnate at best or decline at worst.

4

The Emergence of the Indian Movement

For centuries the Huaorani, "the people," lived relatively undisturbed in the Amazon rain forest in what are now Orellana and Pastaza provinces, caring little if at all that most of the outside world called them Aucas, the Quichua word meaning "savages." The first major intrusion of outsiders came with explorers searching for rubber trees in 1875. Charles Goodyear had invented the vulcanization process in 1839, and the rubber boom reached its peak in the late nineteenth century. The intensive search for the trees in far eastern Ecuador reached its height from 1890 to 1900, and the industry's demise began about 1914.

Petroleum exploration began in Huaorani territory in the 1940s when a Texaco Gulf consortium made the first major finds near Nueva Loja (then named Lago Agrio) in Sucumbíos province in 1967. Dozens of airstrips and hundreds of miles of roads were rapidly built, and a host of new communities sprang up around the oil fields of foreign companies, many of which gave their names to settlements constructed to house the workers. In subsequent decades, the seemingly changeless region underwent the most rapid transformation it had ever experienced. While some observers believed that oil development benefits the Oriente, others considered its impact to be one of cultural destruction and environmental devastation.

The intense exploration and exploitation of foreign companies forced upon the Huaorani ever greater, and potentially devastating, contact with the outside world. Two years after oil was discovered, President José María Velasco Ibarra created a protectorate for the Huaorani. Like many Ecuadoreans, he was desperate to find and exploit oil as rapidly as possible, but at the same time he feared for the safety of oil workers, colonists, and Indians alike. His solution was to give control of the new protectorate to an organization of North American Protestants who had entered the country in 1953, the Summer

Institute of Linguistics (Wycliffe Bible Translators), and charge them with improving Huaorani health and education. The Institute was dedicated to translating the Christian Bible into native languages worldwide; it began working with many Indian groups in the early 1950s, including the Cayapa, Secoya, Huaorani, and Oriente Quichua. Catholic missionaries including the Salesians labored primarily with the Cofán and Shuar. Prior to 1967, the Institute's efforts with the Huaorani were relatively unsuccessful. Five missionaries, for example, flew into the natives' territory, but five days later, on January 3, 1956, they were speared to death. Numerous Indian attacks on oil, timber, and other workers followed. In July 1987, Catholic Bishop Alejandro Labaca and a nun were pinned to the forest floor with twenty-one Huaorani spears as a warning to others to stay out of their territory.

Velasco Ibarra's move in granting the Institute control over the Huaorani protectorate created controversy. Many Ecuadoreans believed that the indigenous people had a right to develop their own cultural tools for dealing with outside penetration and that no one, missionaries included, should impose a response on them. Many more were convinced that the influx of outsiders was irreversible and that oil workers, colonists, soldiers, and others would inevitably introduce deadly diseases that threatened to annihilate the long-isolated and vulnerable natives unless they were inoculated. The indigenous people lacked immunities to the same smallpox, measles, and other diseases that had decimated the Indians throughout the Americas after the Conquest.

Congregating the Huaorani into a few large settlements, the missionaries believed, was the natives' best hope for survival, for achieving a measure of what the Institute deemed progress, and for converting them to Christianity, and they considered the three goals inseparable. The government, for its part, hoped to lessen conflict between the natives and the oil companies by reducing the size of Indian territories and assembling the Huaorani into villages. The Protestant North Americans drew on the Spanish Catholic model of settlements, which five hundred years earlier had been established in the highlands by churchmen. The goals that spanned five centuries were the same: to convert, protect, and civilize the Indians.

Once the natives were assembled, their traditional practices were to be replaced with new ones that focused on schools, churches, farms, and a market economy. Government contracts permitted the Institute to teach the Bible in the schools, where the study of traditional Huaorani culture barely entered the curriculum. Traditional Huaorani education had provided children with the total freedom to learn from adults by watching and imitating them, without either criticism or rewards. Missionary teachers now instructed and judged. By 1990 there were seven Protestant schools in the protectorate.

The impact was enormous. The Indians planted ever-larger fields and became accustomed to a wider variety of crops. Many of them learned to speak, dress, and eat like westerners, and their attraction to new foods, clothing, tools, radios, and other innovations led to estrangement from their traditional forest life of hunting, gathering, and slash-and-burn agriculture. To support the Institute's efforts, the Indians contributed a portion of their labor and crops. Some even took low-level jobs with the oil companies.

After a decade of rapid and profound change for the Huaorani and other indigenous groups, nationalist fervor against the Protestant foreigners emerged during the 1970s, and the Institute was forced to leave the country in 1981. The missionaries had enjoyed a measure of success in their efforts, and in 1980 a cadre of converts formed their own organization, the Ecuadorean Federation of Evangelical Indians (FEINE), which five years later became part of the national Indian movement led by the larger organization, the Confederation of Indigenous Nationalities of Ecuador (CONAIE).

The Institute's evangelists were not the only ones active in the Amazon. International environmental and human rights groups, all concerned with the worldwide destruction of both nature and ancient cultures, were well aware of the changes being wrought in the Ecuadorean jungle. Their exertions on behalf of the natives and the environment evolved slowly, along with the growing recognition of the importance of the Amazon to the world as one of its vital oxygen producers.

The Indians of the Oriente gave major impetus to forming a national Indian organization. In contrast to highland Indians, who for centuries had been isolated and exploited by white hacendados, those of the Amazon had been largely left alone. The Shuar lived in the southeast between the Andes to the west, the Makuma River on the east, the Pastaza River on the north, and Peru in the southeast, in the provinces of Pastaza, Moronoa Santiago, and Zamora Chinchipe. Colonization from the highlands, as opposed to the oil companies, generated the first federation among the long-isolated warrior nationality known as the Shuar or Jívaro: the Federation of Shuar Centers, founded in 1964 with the encouragement and support of Catholic Salesian missionaries. The Salesian Order was formed in 1845 for missionary and educational work and entered Morona Santiago province in 1937.

The Shuar in the mid-1960s numbered around 30,000, and their federation worked to protect native lands from colonists who came in ever-increasing numbers after 1960, when a road linking Cuenca to the Upano Valley was completed. The influx increased dramatically after the 1964 agrarian reform law that was designed to encourage colonization of the Oriente as a means of relieving pressure on land in the highlands, and during the 1960s colonists began to outnumber the Shuar in their native region. With radio programs, a printing press, and other means, the federation defended their way of life from outside intrusion and demanded a pluralist state with a large degree of political and economic self-determination for various Indian nationalities. Although the Salesians were vital in initiating the federation—in part out of fear that the Shuar would be pushed deeper into the forest and become harder to convert unless they were organized—by 1969 their role was reduced. The Shuar soon served as a model for indigenous organizations throughout the region and beyond.

In 1980 the determination of several nationalities of Oriente Indians to protect their lands and cultures resulted in their founding the Confederation of the Indigenous People of the Ecuadorian Amazon (CONFENIAE), which was legally recognized by the government four years later. Every two years it held a congress, which was its main decision-making body. Its main objectives were *the defense and the legalization of the indigenous territories, the preservation of the ecosystems and of the natural resources. Our desire is to promote the social, political, and economic development of the indigenous communities, respecting and rescuing the cultural identity of each nationality, and for the recognition of their rights to the heart of the Ecuadorean state.*[1]

Cultural assertion and environmental protection accelerated in tandem. Following in the footsteps of the Shuar Federation of 1964, provincial organizations sprang up during the 1970s, including the Organization of Indigenous People of Pastaza (OPIP) and the Federation of Indigenous Organizations of the Napo (FOIN). In 1980 efforts were launched to bring all these groups together through the creation of the National Coordinating Council of the Indigenous Nationalities of Ecuador (CONACINIE). The leaders of the Oriente's CONFENIAE and the highlands' Ecuador Ruñacunapac Riccharimui (ECUARUNARI), formed in 1972, forged the new organization. Shortly thereafter, the Coordinator of Indigenous Organizations of the Coast of Ecuador (COICE) appeared, so that all three parts of the country were represented. Years of discussions among the regional groups followed; the terms and objectives of a national indigenous movement were debated and clarified. Although languages and different interests separated them, they coalesced around issues of land use and ownership, ethnic discrimination, poverty, bilingual education, and the need to be heard at all levels of government.

ECUARUNARI, CONFENIAE, and COICE allied in 1986 to form CONAIE. Typical of their layered structure was that of the Organization of Indian Communities of Pastaza, which began locally under that name, had its own autonomous leaders, and was a part of CONFENIAE, the regional entity that in turn was a component of the national CONAIE. Before the century was out, CONAIE was poised to change the face of Ecuadorean politics. After centuries of abuse and neglect, the Indians demanded a voice.

And to some extent they were given one. In 1990 the government granted the Huaorani 2,600 square miles of land, but this amount was only one-third of what they deemed they were entitled to. The legislation making the grant contained the proviso that the natives could not impede the extraction of oil or of any other resource. Predictably, conflict immediately arose among the Huaorani, the oil companies, and the government over land use and environmental protection that still continues into this century.

Tourism also threatened the oil industry. From 1987 to century's end, the number of visitors to the Amazon grew by an average of 12 percent per year, and by 2000 tourism was Ecuador's fourth-largest source of foreign exchange. The government recognized the lucrative link between environmental protection and tourism, and in 1990 the Tribunal of Constitutional Guarantees made it illegal to extract oil in a national park. Higher economics easily trumped tourism and the protection of nature. Foreign oil companies reacted to the tribunal's ruling with a freeze on further investment and exploration. President Rodrigo Borja (1988–1992) reassessed the state's position because his administration depended on oil revenue that dropped in price virtually every year from a high of $35.26 per barrel in 1980 to a low of $16.22 in 1989. Faced with the threat of foreign oil companies cutting off expansion, the president abruptly reversed the environmental policy.

The most controversial area for oil exploration was Yasuni National Park in Orellana province, covering 2,700 square miles of Huaorani land. UNESCO had declared it a "biosphere reserve" because of its variety of wildlife. In 1989, with funds from the Conoco oil corporation, the Ecuadorean government drafted a management plan for the park: one-half of the land was zoned for use by the oil companies, and the other half for the Huaorani. Most environmental and indigenous groups roundly condemned the proposal, and a few months after it was presented the U.S. Agency for International Development selected Wildlife Conservation International to administer an alternative plan: a $15-million program of park management, tourism, and what the agency called "sustainable resource extraction." The substitute scheme won state approval, and in 1993 the Huaorani were incorporated into the park's administration. They worked as guards in part of their homelands, with the oil companies as nearby neighbors. Analysts saw it as a clear compromise of conflicting interests.

In the North, other ethnic groups faced similar pressures. The Cofán people in western Sucumbíos province were a small but thriving nation when Texaco Gulf opened its first commercial well on their land in 1972; twenty years later Cofán culture had all but disappeared. Missionaries during the 1960s feared for the natives' continued existence and asked the government to recognize at least some of their land claims. The grant gave title to 9,000 acres, only a small portion of the territory claimed. Texaco Gulf immediately proceeded to plow a road through their reduced territory, which profoundly altered Cofán culture. As an economic alternative to becoming oil workers, the six hundred native survivors devised a program of "ecotourism" that they themselves operated. Sixty miles of hiking trails, replete with huts, were constructed across the river from where the Indians themselves lived. Tourists began to arrive in 1993 and have continued to come in ever-greater numbers.

Texaco Gulf, the major foreign oil-producing consortium, shipped 1.4 billion barrels of oil, or 88 percent of the total sent to market, over the Andes from 1972 until 1989, when state-owned and -operated Petroecuador took control of the Trans-Ecuadorean Pipeline (SOTE). The Quito government had few environmental regulations for either the pipeline or for oil production in general and made virtually no attempt to assess the industry's environmental impact. With the price of crude in decline from 1981 until the end of the century, the government issued concessions to companies to search and drill for oil in order to hold revenues steady at best and prevent a nose dive of prices at worst. A comparison of per barrel prices and the value of oil exports during the 1980s illustrates the reason for as well as the extent of the expansion. With the price of crude at $32.84 per barrel in 1982 and at $20.32 in 1990, the republic received equivalent amounts of oil revenue in constant 1990 dollars in both years: $1,402 million in 1982 and $1,408 million in 1990.

From the state's standpoint, production increases were absolutely essential to maintain revenue in the face of falling unit prices. In 1993 the investment law of President Sixto Durán Ballén greatly expanded the scope for oil exploration. During the 1994 round of concessions that followed, six territories were auctioned off to foreign oil companies, and in 1996 further auctions awarded additional contracts. Vast areas of the Amazon, including Indian territories and national parks, were opened to the petroleum industry. The authority for the government's actions was an old Spanish principle, enshrined in Ecuador's constitution, that while the land belongs to the people who live on it, the minerals under the ground are the state's. Indian protests aside, successive administrations made the same decision: to expand oil exploration and production in the Amazon.

The process of expansion caused deforestation and contamination in a large portion of the Oriente. Little was done to curtail the damage; oil and the land on which it was found were seen primarily as sources of national wealth. Oil companies, moreover, built roads through the jungle to make the region more accessible to land-hungry colonists (Colonos) from the coast and highlands, and the new arrivals toppled trees to clear fields for cattle and crops. The displacement of the local Indian population continued apace. Although violence flared between colonists and indigenous communities, notably in Napo province in 1985, the influx of land-hungry immigrants proceeded.

One of the world's worst oil spills occurred in March 1987 when an intense earthquake on the eastern slopes of the Andes unleashed waves of soil and rock. The quake caused damage in the provinces of Pichincha, Imbabura, Carchi, and Napo and swept away twenty-five miles of the Trans-Ecuadorean pipeline. The economy's main oil artery was severed, and an environmental disaster ensued. Most Ecuadoreans were more concerned with the economic impact of the catastrophe than with its environmental consequences. But the indigenous people living nearby saw the spill as the most dramatic and disturbing in a long series of calamities that had contaminated soil, lakes, and rivers. By 1989 the pipeline had ruptured at least twenty-seven times in two decades, spilling 16.8 million gallons of crude oil into a delicate web of water. Little of the mess was cleaned up. (The *Exxon Valdez*, by contrast, spilled 10.8 million gallons of oil off the coast of Alaska.)

North American Judith Kimerling in 1989 revealed the nature and extent of environmental devastation from a combination of continual expansion by oil companies and poor environmental regulations. Her findings fostered assistance for indigenous groups. With support from environmentalists and human rights organizations, the Huaorani and others gradually created awareness, at home and abroad, of what they deemed a cultural and environmental catastrophe. Pressure was put on the government and the oil companies to act responsibly by a variety of international environmental groups. The Energy Ministry eventually adopted an environmental protection policy, and in 1990 the oil companies agreed to abide by its rules and regulations. Theory and practice, however, continued on their separate ways as companies continued to seek oil and governments wanting to swell state coffers wished them well in expanding production.

An investigation in 1989 revealed that Atlantic Richfield, a U.S. enterprise, built 1,368 helicopter landing sites and carried out almost 1,000 explosions in its search for oil. In the process it transformed, and some claimed devastated, over 2,500 acres of jungle. Oil spilled from waste pits, and toxic metals found in them caused damage. The contaminants leaked into rivers and streams, and health workers linked the toxins to exceptionally high regional rates of spontaneous abortion, neurological disorders, birth defects, cancer, and other maladies.

Other companies besides Atlantic Richfield were criticized. Natives and environmentalists in November 1993 filed a $1.5 billion lawsuit in the U.S. District Court in New York against Texaco Gulf. The litigants claimed that the company had been negligent in its operations. They charged that during its twenty-six-year contract with Ecuador to produce oil—a lease that expired in 1992—Atlantic Richfield opened 339 wells, cut 18,000 miles of seismic trail and nearly 300 miles of road, and built more than 600 toxic waste pits that discharged large quantities of hazardous waste into the soil and water. Developers took only superficial measures at most to minimize spills and contamination. Ecuador's attorney general, Milton Alava, promised that his country would abide by the U.S. court's decision. The lawsuit failed; there was no compensation for environmental or social damage to the indigenous people or to other Ecuadoreans.

Oil developers were not the only threat to Ecuador's 50,000 square miles of Amazon tropical rain forest, a little less than half the country's total surface area. Twenty percent of this region was designated as national parks or game reserves by 2000, and in theory no logging, hunting, or fishing was permitted. Yet in the three preceding decades, the Oriente suffered serious deforestation, most it caused by colonists, the majority of whom came from the highlands in search of land and livelihood. The influx was in part spontaneous but was also in part the result of deliberate state policy.

Highland colonists clearly contributed to the damage. Following the land reform law of 1964, national administrators tried to reduce frustration over land shortages and unemployment in the highlands by encouraging migration to the eastern jungle where agricultural potential seemed unlimited. Government leaders also believed that colonizers would serve as effective barriers against further Peruvian aggression and therefore gave land to anyone who would cultivate it or raise livestock. Engaging in either activity led to deforestation. And although the green, lush nature of the Oriente made it appear fertile for agriculture, the thin layer of topsoil was exhausted within a few years and the land was left suitable only for pasture. The roads built by oil companies encouraged migration, and during the 1970s and 1980s almost 145,000 people settled in the Amazon area. By 1990 they formed more than one-third of the region's population.

From the mid-1970s onward it grew increasingly clear that serious environmental problems were developing in the eastern jungle, and a number of groups arose to address them. By 1993 at least sixty-one environmental organizations had registered with the government. Some, such as the Nature Foundation, specialized in promoting environmental education in the schools, while others, such as Ecological Science, were primarily concerned with research and the legal aspects of environmental control. By the early 1990s their cause was well established and respected, and in 1994 one of the more prominent environmental groups, Rainbow, received an award from the United Nations in recognition of its efforts to conserve a threatened forest in Loja province. Observers noted that many of the environmental and human rights efforts had been made by women, the so-called Ecochicas, of which Esperanza Martínez of Quito was one of the most active on behalf of the Huaorani. Government interest in the environment proceeded as well. The President's Environmental Commission (CAAM), for example, was created in 1994 with the mandate of devising and supervising national policy.

As the 1990s opened, the Indians of the Oriente and the highlands had ongoing grievances. Such had always been the case, but now they had new organizations through which to bring their influence to bear and to have their complaints heard. Reflecting in 1994 on the origins of indigenous organization in the Amazon, CONAIE's President Luis Macas made connections between environmental destruction and cultural survival:

> *In the Amazon region, there is a crisis caused by the presence of oil and mining companies and their violations of indigenous peoples' rights. The displacement of people from their homes had made it impossible for indigenous people to meet basic living conditions. The oil companies have not only caused the decomposition of our communities and the decomposition of our culture but also the destruction of the ecology. The fight for land is*

> *thus extended to the struggle for maintaining the ecology. . . . Besides provoking a disappearance of species, there has also been a decomposition of communities in the Amazon. Texaco poisoned the places where people lived and worked and threw away its wastes in a totally irresponsible way. . . . What we are really talking about is the extinction of a people.*[2]

Ecuador's Indian peoples have always lived primarily in the middle of the country, but they have never stood at the center of either economic or political power. Constituting 30 to 40 percent of the population according to indigenous leaders, and around 16 percent in the estimation of others, during the closing decades of the twentieth century they grew determined, in a break with the past, to be treated as first-class citizens. They wanted economic and educational opportunities, adequate health care, respect for their cultures and languages, and a voice in national affairs. Once organized into a powerful force to realize those objectives, they engaged in the debate over what kind of modern society Ecuador would have.

Throughout the centuries, land and labor issues caused Indian protests and violence. During the colonial period, uprisings against Spanish efforts at cultural domination as well as revolts against land appropriation, labor and tribute exploitation, and sexual abuse of women were isolated and brutally suppressed by authorities. Animals belonging to Indians were routinely taken for food and transport during civil wars and insurrections, with little if any compensation to their owners. During the nineteenth century, numerous uprisings challenged landowners, among the most notable of which took place in Chimborazo province in 1871, when Fernando Daquilema led a bloody rebellion against abuses in diezmo collection and forced labor for road construction. His and other revolts were harshly quelled by army troops from the nearest garrison. The same pattern of dispersed, disconnected efforts against socioeconomic abuses and discrimination continued until early in the twentieth century.

Intellectual antecedents of the contemporary Indian movement include Pío Jaramillo Alvarado, who published *El indio ecuatoriano* in 1922. Earlier writers, he lamented, had only theoretical as opposed to pragmatic interests in indigenous people. Juan León Mera, for example, who had penned the words to the national anthem, wrote the novel *Cumandá* about idealized, exotic Indians during the 1860s and 1870s; it was strongly influenced by the French romantics. Jaramillo Alvarado sought to change that focus when he documented the exploitation of rural labor and concluded that social justice was a precondition for national economic development. Following Jaramillo Alvarado in writing about deplorable treatment were Luis Monsalve Pozo, Ángel

Modesto Paredes, Gonzalo Rubio Orbe, José de la Cuadra, and Segundo B. Maiguashca. An Indian attorney, Maiguaschca in 1949 wrote *El indio, cerebro y corazón de América*, in which he appealed to the government to return land taken from Indian communities and to help them establish their own schools and training centers. He rejected the notion that Indians could be "modernized" through contact with whites. Instead, he urged pride in ethnic heritage and appealed to Indians to organize in an attempt to achieve social and economic justice. In addition to these and other social scientists, novelist Jorge Icaza published *Huasipungo* in 1934, a scathing indictment of the exploitative hacienda system that graphically depicted the plight of rural Indians. Those Indians living on the rural estates worked four or more days per week in return for small plots of land they were allowed to cultivate for themselves, and indigenous communities close by provided additional labor to hacendados in return for access to pastures, water, and roads.

Indigenous people created one of their first unions in the highlands, with the outside support of socialists, Communists, and the urban labor union, the Confederation of Ecuadorean Workers (CTE). They formed El Inca in Pesillo in 1927. Shortly afterward, Indians formed Bread and Land of Chimba and Free Land of Moyurco, both in the Cayambe region of Pichincha province. Most members were hacienda workers, and their non-Indian leaders emphasized a class struggle within a national context. Organizers, it appears, gave particular attention to those working on state haciendas, lands that had been taken from the Church by the Ley de Manos Muertas during the Liberal Revolution early in the century. Renters administered the lands and employed indigenous workers. The preservation of Indian culture, language, communities, and traditional territory held no interest for these organizers, who simply assumed the need for Indian integration into national society. Class struggle was stressed to the near exclusion of ethnicity and cultural concerns. Later, particularly from the 1980s forward, the Indian movement emphasized both as well as cultural needs. When the various rural unions tried to form a united movement in 1931, government repression thwarted their efforts. Less than two decades later, the Ecuadorean Indigenous Federation (FEI) appeared in the highlands in August 1944, with the support of the urban-based Ecuadorean Workers Confederation and the Communist Party. Party members who were not indigenous people led the FEI. It urged land reform as it tried to organize indentured peasant demands and sought the parceling of large estates, fewer working hours, better salaries for laborers, and ending the huasipungo system.

By 2000 the Indians had a major political movement, CONAIE. Assembled through twenty-six regional groups into a national organization, they spoke with a confidence and pride scarcely seen thirty years earlier. CONAIE has come to play a central role in national affairs. The movement recognized the close relationship between those who held political power and those who maintained the social and economic structures that for centuries had kept the masses of Indians poor and powerless. Some saw it as a divisive threat, as illustrated by the derisive assertion of Interior Minister Francisco Huerta, who said in late January 2000: *We are not going to cure the republic with witchcraft or with hostile protests against those who are not wearing ponchos.*[3] The Indians were not deterred; they took pride in the poncho, which had become a symbol of ethnic identity.

Not all politicians shared Huerta's sentiments. Shortly after assuming Ecuador's presidency in January 2000, Gustavo Noboa, whom the Indian movement had opposed, spoke of old injustices: *It is indispensable that Ecuador has peace, but to have peace you need freedom, and to have freedom you need justice. And the Indian population needs justice. The indigenous theme is a long-standing one. The fact is that they have been deceived for centuries, and their demands are right in part.*[4] Moreover, Leslie Alexander, U.S. ambassador to Quito for a time during the 1990s, was appalled by the status and treatment of the Indian population. Why, he asked, was Ecuador's political leadership *incapable of producing wealth for the benefit of all*? And why were the Indians, in the eyes of so many, *perceived as unequal because of their blood*?[5]

The isolation of rural communities and the domination of land and Indian labor by large estate owners made organization difficult. Nonetheless, a wave of social unrest swept rural regions during the early 1960s. Some 2,000 Indians, for example, rioted in 1961 on a large hacienda in Chimborazo, the province with the greatest concentration of indigenous people, and three years later another group seized a hacienda in Cayambe in northern Pichincha province. CONAIE's origins and development as a political movement go back to these and other conflicts and to a number of now-defunct groups that emerged in response to the unrest. The most important to develop in the highlands was the National Federation of Farmers' Organizations (FENOC), formed in 1965, that included both Cholos and Indians and that operated on a labor union model without ethnic dimensions. The military regimes favored FENOC over the Communists, whom they suppressed, and over other leftist unions active in the countryside. FENOC, they thought was the least radical and threatening of the organizing efforts and would help implement agrarian reform. The driving force for the origin of FENOC and other groups was the quest for land, ever central to the indigenous population.

Agrarian reform was enacted in 1964, spurred in Ecuador as elsewhere in Latin America by the same forces that fostered indigenous organizing: growing discontent with exploitation and poverty within the Indian community and among more of the outside world, Washington's Alliance for Progress that offered money in return for a measure of reform, the new emphasis of the Catholic Church on social justice after Vatican II, a growing number of rural workers' organizations, and the agitation of a growing group of well-educated indigenous leaders.

Raúl López, the Catholic Bishop of Latacunga in Cotopaxi province, remembered the common newspaper advertisements in the 1950s that, as he put it, *offered haciendas for sale with Indians included, as if they were cattle or horses*.[6] The bishop was referring to huasipungueros, resident Indian laborers. Marxist parties on the far left had long backed agrarian reform, and as early as 1923 the Liberal Party adopted as its principles an end to the archaic institutions of huasipungo and concertaje, and the kind of agrarian reform that would eliminate large and inefficient estates. In 1953 the party reaffirmed these goals. The impulse to modernize the economy, manifest in the urban sector in efforts in industrial import-substitution policies, was reflected in a stronger focus on rural development in Ecuador as well as in many other countries in the region.

Unprecedented population growth that put pressures on the land and increased internal consumption played a major part in motivating rural modernization and restructuring. It took the entire nineteenth century for the population to double, but only the first half of the twentieth for it to double again. After midcentury, however, it started doubling about every three decades. The growth from 2,369,800 Ecuadoreans in 1938 to 4,476,007 in 1962 was significant, but the explosion to 12,400,000 by 2000 placed vastly greater burdens on the same fields and pastures. Ecuador's growing rural population was made more mobile (and one can assume more assertive) by the abolition of imprisonment for debt in 1918. Advancing urbanization shifted the balance of political and economic power away from rural landowners, making agrarian reform far easier in political terms than it was only a few decades earlier. Urban demands mounted alongside rural pressures as the rapid demographic shift unfolded. Whereas Quito's population had grown at 0.4 percent per year from 25,000 in 1780 to 39,600 in 1888, that rate soared to 4.5 percent from 1950 to 1974 and its numbers increased from 210,000 to 355,200. By 2000 there were almost 1.5 million Quitenos. Like those in the countryside, people in the cities organized to demand changes ranging from adequate housing to sanitation services and clean water.

The agrarian legislation placed limits on the maximum size of estates, but its provisions were flexible. Many owners, especially in the northern highlands, sold enough of their acreage to escape the reform, and a large number of them modernized their reduced estates. In the center and south there was less modernization, and hacendados also sold a considerable amount of land to neighboring Indian communities, Cholos, and others. Although many haciendas were divided, lighter-skinned Ecuadoreans kept most of the best land. Indigenous people received the hilliest and least fertile plots, and most continued to labor on the large estates. Former huasipungueros tied to haciendas were the main beneficiaries of the restructuring, as they won title to small plots and saw the termination of the most servile form of labor relations with hacienda owners. The reform did not come easily. Provincial landowners, who perhaps were the most reactionary segment of society, condemned the measure and through provincial associations determined to prevent the law's application.

Between 1964 and 1984, government agents distributed approximately 1.8 million acres to 95,000 families, each of whom received on average about seventeen acres. By 1982, about 25 percent of haciendas larger than 100 hectares had been impacted. The net result was a marked decline in the concentration of land ownership. Whereas large estates monopolized over two-thirds of the land prior to the reform, by 1985 small, medium, and large estates were of roughly equal proportions in the highlands. It was a welcome improvement for those with no or little land, but in many ways the impact was shortlived. Within a few years, shared inheritance resulted in most recipients being left with fewer than twelve acres. Overcultivation followed, and it became increasingly difficult to make a living on the ever-smaller plots. Financial credit and technical help from the government, moreover, were scant at best, and as a consequence of these limitations as well as of population growth and subdivision of parcels, Indian farmers took over more of the steeper slopes in the mountains. In the process they irreparably damaged large stretches of ecologically vulnerable territory.

Land ownership remained highly inequitable, despite the 1964 agrarian reform and its successor measure of 1973. The latter legislation emphasized expropriation of inefficiently cultivated land with its focus on production, not redistribution. An estate could be expropriated if one-fifth of the land was not cultivated, if inefficiently farmed, if worked by squatters without legal title, or if located in an area under great population pressures. The law forced the haciendas of the highlands to undergo a measure of modernizing as the second half of the century progressed and the number of midsized, market-oriented holdings increased. Many people in the northern highlands, for example, turned to dairy production, and the output of milk grew by about 7 percent per year in the 1970s. Nonetheless, in the 1980s only 5 percent of farms exceeded 50 hectares, a number that accounted for over 55 percent of cultivated land. In contrast, 80 percent of farms encompassed fewer than 10 hectares and comprised a mere 15 percent of farmland. At century's end, land distribution remained extremely skewed; 1994 figures showed that 1.6 percent of farms in the highlands controlled 43 percent of the cultivated land. The importance of agricultural employment fell sharply in the twentieth century's last decades as urbanization, manufacturing, and industrialization expanded. Thus, only 30 percent of the workforce was employed in farming or fishing in 1990, a major drop from the more than 60 percent so employed in 1969. Notwithstanding the spectacular rise of the oil industry, about one-half of Ecuador's foreign currency earnings in 2000 came from agriculture and fishing.

Large sectors of the rural population opposed the thrust of the nation's agrarian policy. Luis Macas, a Saraguro from the southern highlands and an early CONAIE president, expressed this view in 1994: *There is currently a plan that they call modernization in the agricultural sector, which is being processed by the Inter-American Development Bank. The plan's goal is to create a system which is run by agro-industry, one that would encourage agricultural production for export, disregarding the basic food needs of the Ecuadorean people.*

Macas, who holds a law degree, also presented the outlines of CONAIE's alternative approach to reforming the countryside:

> *CONAIE's agrarian reform proposal now in the Congress is derived from the way that indigenous people see this question. It is meant to benefit not only the indigenous people but the entire agricultural sector. The first aspect is a restructuring of land ownership. It is impossible to talk about agricultural development when land is in the hands of the very few. The second chapter deals with making production more dynamic and sustainable. The goal is not only to try to meet the needs of the farmer and his community, but the internal needs of all of Ecuador. This part of the law is directed at encouraging sustainable development of the Amazon, but it would also be applied in the coastal region. The last part deals with the democratization of government institutions overseeing the agrarian sector in order to increase the participation of the indigenous peoples and*

> *farmers, because now there is no democratic participation in these institutions. It is all a personal decision of the president.*

Land reform spurred significant growth in indigenous peoples organizing, as communities endeavored to obtain lands they deemed unjustly taken by large property owners. The 1937 Law of Communes (Ley de Comunas) established a legal status for Indian communities when it mandated that they were required to form local governments with town councils (cabildos). State authorities legally recognized these units, and the communities proceeded to seek autonomy over natural resources for local development. Although the 1937 law did not allow huasipungueros to participate in the cabildos, with the elimination of the huasipungo system in 1964 they were free to do just that. Hundreds of cabildos were organized in Indian districts throughout the highlands during the 1970s and 1980s, and these local governments played a prominent role in the development of indigenous organizations.

Macas in 1994 succinctly summed up the relationship between indigenous land issues in the three major parts of the country, and he explained why CONAIE was formed:

> *The problems facing the indigenous peoples are deeply connected to the issue of land ownership. When the colonizers arrived, they cleared out the Indians. Today, land is concentrated in the hands of the few, and many of our people don't have any land. In the Amazon region, there is a crisis caused by the presence of oil and mining companies and their violations of indigenous peoples' rights. The displacement of people from their homes had made it impossible for indigenous people to meet basic living conditions. The oil companies have not only caused the decomposition of our communities and the decomposition of our culture but also the destruction of the ecology. The fight for land is thus extended to the struggle for maintaining the ecology. Land ownership is also the central issue in the highlands, and it is an issue that must be resolved through negotiations. What often happens is the government tells the community that they should try to buy their land from landowners who then put a very high price on this land. In the coastal area the principal problem is the cutting of the forests and the tricking of community leaders into allowing this to occur. The lumber companies are trying to get concessions of large areas to cut down the forests.*[7]

Efforts at Indian unification grew slowly. FENOC, not a strictly indigenous organization, focused on land issues rather than on ethnicity, bilingual education, cultural preservation, racism, and a multicultural, plurinational state. During the 1970s, socialists moved into leadership positions in FENOC and pushed it increasingly to the left. In June 1972 the first effective Indian organization, ECUARUNARI, spread through the highland provinces of Pichincha, Tungurahua, Chimborazo, Cañar, Esmeraldas, and Imbabura. ECUARUNARI, like FENOC, initially focused primarily on land issues and barely dealt with ethnic concerns. The Catholic Church encouraged and aided the group. Later the close affiliation was severed as the movement's leaders concentrated on Indian identity, land issues, and autonomy from oversight of any outside civil or religious organization. ECUARUNARI's evolution away from its religious origins was similar to that of the Shuar Federation, launched in 1965 in the Oriente with assistance from Catholic Salesian missionaries. Many in the Indian movement later viewed Catholic and Protestant sects and state agencies as impediments to unity. ECUARUNARI is recognized by CONAIE as its major antecedent. In addition to CONAIE, in 2000 one of the Indian movement's central organization was FEINE, the Protestant group founded in 1980 and headed by Marco Murillo. With its associations of evangelical Indians in most highland and Oriente provinces, FEINE was strongest in Chimborazo. Founded *to be a guiding institution of spiritual development*, it also sought to *implement programs and projects that will develop and strengthen intercultural relations in Ecuadorean society.*[8] Cultural revitalization, bilingual education, training, and technical development were major concerns along with fostering understanding and acceptance of spiritual, social, ideological, and economic diversity. The other major entity was the National Federation of Rural Campesinos, Indians, and Negroes (FENOCIN) of Pedro de la Cruz, which emerged from and replaced FENOC.

Blanca Chancoso and other leaders claimed that enhancing identity and pride of a people long called ignorant and dirty by outsiders were prerequisites to forging a national Indian organization that could demand economic and political gains. They rejected pleas to ignore ethnic differences and to assimilate, an approach urged by President Rodríguez Lara, who in 1972 argued that there was no longer an Indian problem because the natives became like white men when they accepted the goals of national culture. Part of the challenge to establishing identity was determining the proper terminology. In the 1980s the term "indígena" was adopted, while "aborigines," "natives," and others were considered and rejected. In the future, CONAIE leaders hope that indigenous people will be called by their cultural name (Shuar, Otavaleno et al.) as opposed to the general label "Indígena."

Organizing efforts accelerated as the 1980s unfolded. To recap mobilization in the Amazon, in 1964 the Shuar founded the region's first ethnic federation with the assistance of Salesian missionaries. International human rights and environmental groups were soon attracted to the movement, and cultural assertion, environmental protection, and procuring territorial rights against international oil and mineral companies proceeded together. Other federations followed, and CONFENIAE in the Amazon was established in 1980. Between the formation of the Shuar Federation and the founding of CONFENIAE, in 1972 ECUARUNARI was launched in the highlands and embraced ethnic as well as land issues. A united organization including all regions of the country came next.

During the 1970s and 1980s a number of other groups appeared to advance the interests of Indian communities. Confronted with deteriorating economic and social conditions, nongovernment organizations (NGOs) emerged because many people believed that Quito's political bureaucracy was out of touch with the needs of isolated rural communities. These NGO members sought to bridge the gap created by ineffectual federal agencies. Most of the self-help organizations, whose efforts were imitated by urban dwellers, focused on the special needs of a target group. Many were spontaneous responses of the poor. They received some assistance from religious organizations and international aid agencies in promoting food and other cooperatives and in offering low-cost meals. NGOs also were created for similar purposes in other Latin American countries.

Despite ECUARUNARI, CONFENIAE, and the NGOs, a national indigenous organization emerged slowly. Indians had generally shunned the political system that for centuries barred them from voting through literacy requirements. Many thought their interests were better served by ignoring a process that had long disregarded them; they feared that embracing politics meant losing ethnic identity. Indian communities faced diverse problems and had different regional and other interests. Those in the highlands, for example, struggled for land redistribution, while for those in the Oriente the major challenges were the oil companies, deforestation, and colonists. Despite the obstacles, CONACINIE was formed to bring the disparate groups together through the central unifying themes of ethnic and cultural issues.

The nation's general interest in the Indian population broadened in the 1970s, and the attitudes of many Ecuadoreans changed as natives organized and asserted themselves. In 1978 the military government drew up a new constitution that extended the vote to illiterate adults. The overwhelmingly Indian countryside gained significant political ground with the change. Although by no means all of the nation's illiterates were Indians, in comparison to criollos and Cholos a higher proportion of natives lacked reading and writing skills. In the 1968 presidential election for literates only, 851,359 voted; in the 1979 election, with the new constitution in effect, 1,496,805 cast ballots; and in 1984 the number rose to 2,680,798. In another sign of the growing strength of the indigenous movement, Jaime Roldós made the symbolic gesture, after being sworn in as president, of delivering part of his 1979 inaugural address in Quichua. As Indian organizations matured during the 1980s, many of their leaders commented that the appearance of political parties and government agencies in Indian communities, each with a host of false promises, had the same divisive impact as had the earlier Protestant missionaries.

Following years of organizing efforts, on November 13, 1986, CONAIE was established around issues of land use and ownership, racial discrimination, poverty, bilingual education, and the need to be heard at all levels of government. Two movements joined to form its backbone: ECUARUNARI, the highlands entity established in 1972; and CONFENIAE, the Oriente-based league dating from 1980. Ancient struggles for land and culture in the highlands, and more recent battles against petroleum and other incursions in the Oriente, finally united the vast majority of the nation's Indians. Although the indigenous population of the highlands vastly outnumbered that of the Oriente, the homeland of the latter was vital to the petroleum-dependent nation. Many university graduates took leadership roles, including Blanca Chancoso, Miguel Taukowash, Cristóbal Tapui, Luis Macas, Antonio Vargas, Salvador Quishpe, Nina Pacari, José María Cabascango, Carlos Viteri, Miguel Lluco, and Leonidas Iza. The country's modernization policies during the oil boom of the 1970s helped produce a new generation of well-educated indigenous leaders. During that decade the country underwent an unprecedented expansion of education and a proliferation of government development agencies.

Twenty-six regional affiliate groups, as noted earlier, gathered on a national level and claimed to represent four million indigenous people. Traditional political parties had given too little attention to the Indians, they said, and they now formed their own independent, autonomous organization, CONAIE, which was not affiliated with any party or with any state, foreign, or religious institution. (CONAIE, however, does accept financial support and advice from a variety of human rights, environmental, and other groups from Western Europe and North America.) During the 1990s, CONAIE brought together under its rubric an alliance of entities from the Oriente, the highlands, and the coast. CONFENIAE encompassed such groups as the Organization of Indian Communities of Pastaza; ECUARUNARI assembled such entities as the Federation of Indians and Campesinos of Imbabura (FICI) and the Union of Indian Campesinos of Cotacachi (UNORCAC); and CONAICE sought out federations on the coast. These Indian organizations acted in pragmatic and erratic alliance with urban-based groups, including the Coordinated Social Movements (CMS) headed by socialist Napoleón Saltos, whose mainstays were unions of public employees, the most powerful of whom were in the petroleum industry; and the Popular Front (FP), headed by Luis Villacis in 2000.

The Indian movement demanded civil rights and the end of abuses. César Umajinga, president of the Cotopaxi Indigenous and Peasant Movement in Latacunga, claimed that resentment against centuries-old insults was the major impetus to organization. Looking back from the perspective of 2000, Umajinga recalled: *We were tired of bus drivers insulting indigenous passengers, refusing to allow them to board or telling them to take off their hats. After several of our people died from being thrown off buses, we decided it was time to demand respect and an end to such abuses.* [Indians were] *tired of being on the margin and treated as orphans by the government.*[9]

CONAIE's member federations also coalesced around demands for land reform, territorial and water rights, and what they called a *multinational and multi-ethnic state* in which Indians had the right to teach their native languages in their own schools, to protect and preserve their cultural traditions. The annual Sun Celebration called Inti Raymi and natural medicines and traditional healing practices were promoted by CONAIE. Moreover, in 1989 it signed an agreement with the Ministry of Education that established a national program of bilingual, bicultural education that it designed and managed. CONAIE also promoted traditional forms of land use such as cooperatives, organic farming, natural pest-control management, and trading among Indian nationalities. The vast majority of indigenous farmers, CONAIE leaders pointed out, lived in abject poverty yet made enormous contributions to the nation. In 1995 government statistics set the number of poor people at 56 percent; with an 80 percent poverty rate, the Indians were the poorest of them all. Nonetheless, with only 35 percent of the country's arable land, the Indians managed to produce 75 percent of the country's basic foods.

From 1988 forward, CONAIE's emphasis on ethnic and cultural issues led it to advocate a restructuring of government to ensure more direct political participation by all indigenous nationalities and social groups. Decentralization of the state was deemed essential by its leaders. Luis Macas was fond of quoting Ecuador's most famous president, José María Velasco Ibarra, who once told the Tsachila Indians: *Certainly you are an autonomous people, so you have the right to administrative autonomy.*[10] The multinational society that CONAIE envisioned differed radically from the country's traditional system, which it labeled the Hispanic-Ecuadorean Nationality. CONAIE's leaders called it a fake democracy in which a small elite owned most of the wealth and used its power to draw up a government, constitution, and rules for the perpetuation of its position. With its European principles and procedures, the national system excluded Indians who had evolved their own way of governing within their own territories. It had likewise placed on the margins the Afro-Ecuadorean people of Imbabura and Esmeraldas provinces. Macas summarized the movement's attitude toward so-called development and modernization:

> *Either we definitively abandon this system, so stifling, so aggressive and violent for the indigenous peoples—but not only in economic terms, but also in its definitive ravaging of our knowledge, our cultural values—either it absorbs us or we liberate ourselves. The application of neoliberalism is quite well designed to finish with everything—the disappearance of centuries-old cultures, the disappearance of peoples, of life itself. I don't know how to interpret this term modernization. For us, modernization obviously is the changing of structures that don't currently serve to advance the economic, social, and cultural development of peoples, to try to realize at least some of the fundamental elements that are components of the development of the humanity that we currently live. . . . I have only seen a single dimension here, which is the economic part. And we know clearly that this comes from the interests of the sectors that have always hegemonized economic power, and therefore political power. They try to monopolize in their hands all the resources that in one way or another are also the conquest of our peoples. For example, here in Ecuador what they want is to actually confuse the terms modernization and privatization, that is, to use the label modernization to privatize the state's resources, which are resources that belong to all Ecuadoreans.*[11]

CONAIE's leaders insisted that their goal was not the simple taking of power. Rather, they intended to transform a structure that was exclusive, antidemocratic, and repressive and replace it with another for the purpose of ending centuries of oppression, misery, and poverty. Eurocentrism and ethnocentrism, they argued, eventually had to give way to a new economic order and a multinational state. CONAIE invoked the memory of indigenous leaders of the past back to Rumiñahui and concluded that the struggle of the twentieth century was the same as it had been five centuries earlier—a fight for a measure of independence from all conquerors, old or new. The seven colors of the rainbow on CONAIE's banner, or huipala, symbolized the unity of the indigenous people. The military and others saw it as the notion of a state within a state, an affront to the Ecuadorean national flag.

CONAIE was organized from the bottom up. At its base on the canton and parish levels were cabildos in the communities. Provincial federations were formed by unions of cabildos, which in turn were grouped together in regional confederations. The Oriente, highlands, and coastal confederations were the strata immediately below the national level. CONAIE leaders continually stressed the importance of having issues discussed and decisions made from the community level up. The movement's structure as well as its success were influenced by the weakening of the traditional rural system that for centuries had been dominated by a handful of large hacienda owners, the gradual modernization of agriculture, and the persisting presence of Indian communities and cabildos—entities that made up CONAIE's base. To enter the national entity, one first had to be a member of one of the smaller groups on the organizational ladder. In June 2000, Estuardo Remache, newly elected president of ECUARUNARI, explained that for an Indian community to join a parish organization of his confederation, and later CONAIE, it had to have two prior years of active participation in mingas (community service) in constructing parish schools, roads, irrigation ditches, or similar activities. The bottom-up system, many thought, ensured a continued supply of capable and experienced leaders. Local communities were able to make their own decisions on whether to follow CONAIE's calls to action.

Every six months each provincial federation called an assembly, which consisted of the directors elected by the congress and of nine representatives. The congress, elected by the federation, convened every three years to elect nine directors to a council of government and to establish political and social directions for the organization to follow. The council executed the acts of the congress and the assembly. The president of CONAIE acted as yet another executive. Other components show that it has become a parallel governmental organization with outreach efforts, both within the country and abroad. In November 1999, CONAIE's congress elected members of the council: Antonio Vargas, president; Ricardo Ulcuango, vice president; Blanca Chancoso, director of International Relations; Rosa Alvarado, director of Health; Anibal Piaguaje, director of Youth; Ángel Gende, director of Territories; Ruth Peñafiel, director of Women; Daniel Tigre, director of Education; and Calixto Napa, director of Fortifying Nationalities. Chancoso said of the movement's structure that *all major resolutions are deliberated in the communities, and from there they pass to the next level. Because of that we are a little slow in making important decisions, but it is worth doing it that way because here individualism does not fit—we travel together.*[12]

Thirteen Indian nationalities united in CONAIE in 2000. From the coast and the confederation of COICE were the Chachi, Tsachila, and Awá nationalities; the Manta and Huancavelica people were awaiting entry. In the highlands and forming the ECUARUNARI confederation were thirteen Quichua peoples: the Saraguros (Loja and Zamora provinces), Cañaris (Cañar province), Puruháes (Chimborazo province), Warankas (Bolívar province), Panzaleos (Cotopaxi province), Salasacans, Chibueloes, and Quishapinchas (Tungurahua province), Quitus (Pichincha province), and Otavalos, Natabuelas, Karamkis, and Kayampis (Imbabura province). In the Amazon, the confederation of CONFENIAE had the Shuar, Achuar, Siona, Secoya, Cofánes, Huaorani, and Záparo, and organizations in Pastaza, Napo, Orellana, and Sucumbíos provinces.

CONAIE has never been an exclusive movement. From the outset it joined labor unions and other national organizations to protest government policies it deemed contrary to Indian interests. In June 1988, for example, it took part with the urban-based United Workers Front (FUT) in a national one-day strike aimed an increase in the minimum wage and a freeze in the prices of basic necessities. Some divisions within CONAIE have arisen over the degree to which it should be a separatist ethnic movement focusing on indigenous issues alone as opposed to a broader entity dealing with the diverse concerns that affect all Ecuadoreans. There was also divisiveness over separatist notions of political and economic structures. In addition, a breech is seen between those who want to lead with an iron hand with the objective of eventually dominating the entire nation, and those who respect diversity and seek a pluralistic, multicultural, and multinationality country.

Under the leadership of Luis Macas, CONAIE came to national prominence two years later, during the June 1990 Levantamiento Indígena (Indian uprising). The Levantamiento was sparked by events in the Oriente. The Indian mobilization stunned and surprised the country and demonstrated that CONAIE was to be taken seriously—a force that was committed to resolving centuries-old ethnic and economic conflicts. President Rodrigo Borja, confronted with oil prices about one-quarter below those of the previous administration, had given foreign oil companies new concessions to drill in 3.5 million acres in the Amazon. This was the homeland of the Huaorani, who were not consulted. Outraged, they marched on Quito. Their opposition to the oil companies' expansion was backed and broadened by a wide range of labor, environmental, and other organizations. Thousands of CONAIE members descended on Quito in June 1990 not only to support the Huaorani demands but also to push the claims of highland Indians for the resolution of land disputes. They congregated in El Arbolito park. Many who did not go to Quito kept their agricultural produce off the market, obstructed the nation's principal roads, and participated in mass marches to regional capitals.

Arriving from all over the country, CONAIE members peacefully took over Santo Domingo Church and blocked the Pan American Highway, the highlands' most important north-south artery. The disruption ended when the government agreed to negotiate with CONAIE on sixteen issues presented by the movement. The indigenous people, CONAIE's leaders believed, were now protagonists and not to be taken for granted. The organization had gained official legitimacy through state recognition.

CONAIE again came to Quito about a year later, in May 1991. This time 100 of its members occupied Congress to demand that Ecuador be constitutionally considered multiethnic and multicultural and to insist on amnesty for 1,000 peasants charged with offenses for their conduct during recent land disputes. Luis Macas warned that if the demands were not met, the Indians would form their own parliament and government. Ignacio Pérez, president of Ecuador's Chamber of Agriculture, accused the Indians of attempting to form a state within a state and claimed they would do away with the legitimate property of landowners. The leader of the legislature, Edelberto Bonilla, nevertheless agreed to include CONAIE's demands in the agenda of forthcoming sessions, although doing so accomplished little: the oil companies kept their concessions, land reform barely progressed, and most of CONAIE's other requests went unheeded. Still, the movement's protests strengthened indigenous self-confidence and bought it national attention, which helped to ensure that its later demands would be better heard. Leaders of the demonstrations, among them Luis Macas and Nina Pacari, became national figures. In 1996 both individuals were elected to Congress from the newly formed Pachakutik (New Awakening and Revolutionary Change) party.

Ecuador's Indians were in the forefront of a hemisphere-wide movement. CONAIE's status was recognized in 1990 when 300 delegates from twenty countries came to Quito to participate in the four-day Continental Indian Congress. That same year Indian protests paralyzed the highlands for several days; and a little over two years later, on Columbus Day 1992, 20,000 Indians from all corners of the country descended on Quito to stage another mass demonstration. Once again the protest supported the Huaorani stance against the oil companies' intrusion onto their lands and favored cultural autonomy. Buses hired by female environmentalists from Quito, the Ecochicas, helped transport natives from the Oriente to the capital. About half of the protesters made it through a cordon of military barriers erected around the city to keep them out. They gathered in Independence Plaza in front of the presidential palace to hear speeches condemning five hundred years of oppression by the Spanish conquerors and their descendants. CONAIE developed a full slate of demands: that 1 percent of the cost per barrel of oil pumped in the territories of indigenous peoples be used to benefit the natives, that the constitution recognize the multinational and multicultural character of the country, that amnesty be given to those sentenced as a result of their struggle for land, that a fund be created for the resolution of land conflicts, that another fund be established for indigenous development programs, and that scholarships be made available to support the training of Indian students. Quito's leading newspaper, *El Comercio*, reported that the Huaorani *captured the attention of the entire nation.*[13]

The Huaorani met the president. During the demonstrations their leaders conferred in Carondelet with Durán Ballén. They told him that he should recognize their fundamental right to prevent their forest homeland and its rivers from being polluted by oil companies. They asked him why he did not allow anyone to toss garbage into his own home, yet he permitted the oil industry to throw waste into theirs. Moreover, they informed him, they did not want to exchange their way of life for his, even for schools and airplanes. Indeed, if he thought the oilmen offered a good life, fine, but he should give them his home instead of theirs. Not surprisingly, little changed after the dialogue with Durán Ballén. Ecuador's expenditures exceeded revenues year after year, the price of oil was stagnant after a major decline in 1986, and, like those presidents who came before and after him, Durán Ballén deemed it essential to encourage oilmen to develop all of the oil fields they could find.

Many observers noted that as political parties on the left and labor unions declined in membership and strength—due in part to the antidemocratic policies of the military regimes from 1970 to 1979 and ideological divisions following the Sino-Soviet split and the Cuban Revolution of 1959—Indian organizations advanced in numbers and might. In 1963 the Ecuadorean Socialist Party (PSE), founded in 1926, divided when a group determined to copy Fidel Castro established the Ecuadorean Revolutionary Socialist Party (PSRE). The following year the Ecuadorian Communist Party (PCE), united since 1931, fractured into pro-Soviet and pro-Chinese factions. Labor splintered as well. The Confederation of Ecuadorean Workers (CTE), founded in 1944, was wracked by warring between Communists and socialists, and the Ecuadorean Confederation of Catholic Workers (CEDOC), established in 1938, underwent a schism between Church-influenced moderates and militant socialists.

The FUT had been founded in 1971 with the aim of uniting most unions. Few people were surprised when on May Day 1993 the federation's call for a nationwide strike against President Durán Ballén's economic policies had a negligible impact. Three other calls for national shutdowns from 1984 to 1988 went virtually unnoticed. A few weeks later, when CONAIE joined the cause, the effort affected the entire country. In 1994, CONAIE mobilized against Durán Ballén's proposal to amend the agrarian reform law to allow the sale of unused and communal Indian land rather than simply give titles to poor farmers. The president hoped to modernize agricultural production, but CONAIE saw his planned legislation as a further effort to alienate native fields. Indigenous protests stopped the president's project, but only after Indians together with Luis Macas marched on Congress, where they were met with tear gas and clubs. In 1995, yet again, the movement mobilized against Durán Ballén's eleven plebiscite proposals that dealt with privatization and labor relations. Along with the Coordinated Social Movements (CSM), made up of unions, students, women's organizations, and others, CONAIE defeated every one of the president's offerings that they deemed contrary to the interests of the nation's poor.

The founding of the political party Pachakutik in 1996 marked the Indian movement's entrance into Ecuador's traditional electoral process. The party was not formally allied with CONAIE but remained a separate, independent organization. CONAIE's president, Antonio Vargas explained: *I am not going to enter a political party. That way, with uprisings, we are in a better position, without bitterness and resentments, to fight for our people.* In addition to indigenous people, from the outset Pachakutik included others on the margin of society, among them women, ecologists, human rights activists, and a number of NGOs. *We decided that we wouldn't just fight for our issues,* said prominent member Nina Pacari. *We decided to fight for issues that affected all of society.* She recalled that before 1996 only one or two indigenous Ecuadoreans had ever served in Congress, whereas in January 2000, six Pachakutik members held places in the 123-seat legislature. Four of them were Indians. The party's emergence allowed the indigenous movement to focus on national concerns such as low wages, corruption, and privatization and to forge a broader base of support. That concentration was criticized from the outset by many in the movement who thought that a focus on indigenous problems should take priority over general or national issues.

Despite Pachakutik's modest electoral victories in Congress, in 2000 CONAIE's Vargas doubted that traditional Ecuadorean elections, unless drastically reformed, would be adequate for pursuing the Indian movement's goals. He explained: *Those who have managed the country are the rich and powerful. They have the press and economic resources which the people do not possess. The participation of the people is restricted. Almost all of the laws benefit only a certain group. It is necessary to democratize the system to have a larger participation of all sectors.* [14] And despite having achieved a much higher profile for indigenous issues, some valuable propaganda victories against foreign oil companies, and occasional successes in modifying government policies such as price increases and agrarian restructuring proposals, CONAIE's leaders believed that they had failed to gain adequate improvements for the Indian people.

A major accomplishment came in the 1998 constitution. Using both politics and protest, a number of native rights and protections were written into the charter, which explicitly

> *recognizes the existence of Indian and Negro communities, defined as Indigenous and Afro-Ecuadorean nationalities . . . recognizes the traditional forms of organization, authority, and the exercise of that authority . . . recognizes the right of Indians to resolve social conflicts with their own norms, without separating from ordinary justice. The two should produce harmony . . . recognizes Indian medicine . . . recognizes the system of intercultural and bilingual education . . . recognizes the community intellectual property of the ancestral knowledge . . . recognizes the right to conserve their lands and not to be displaced from them . . . recognizes the legality of and guarantees the possession of ancestral Indian lands . . . recognizes the right to use, administer, and conserve its own renewable natural resources . . . in case the natural resources are not renewable, it recognizes the right to be consulted before exploration and exploitation. Furthermore, they should be the beneficiaries of those projects. And if there is environmental damage, they should be compensated.* [15]

The constitution incorporated as duties of all citizens the three basic principles often referred to by CONAIE leaders: *Ama Llulla, Ama Shua, Ama Quilla*: Do not rob, do not lie, and do not be lazy. These phrases, which had been common during the Inca empire, were frequently used in the nation's political discourse.

From the outset, education was a major issue for the indigenous movement. National gains against illiteracy had been significant since 1950, when the rate stood at 44 percent. By 1990 the level of illiteracy had declined to 14 percent. In urban districts and among men the percentage of illiterates was much lower than the national average, but nonetheless in those provinces with sizeable Indian populations—namely, Chimborazo, Cotopaxi, Bolívar, and Cañar—the illiteracy rate was twice the nation's norm. In theory each child was entitled to nine years of compulsory education, but government figures revealed that in practice most attended school for only six years when poor peasant families found the costs of registration, uniforms, and reading materials prohibitive. During the 1990s the growth of the number of students in school stagnated as a consequence of government cutbacks and the rising rate of poverty. Although the constitution allocated 30 percent of state spending to education, the goal was never reached, and in 1994 expenditures for this purpose accounted for only 18 percent of the budget. Bilingual education programs for Quichua and other Indian-language speakers, moreover, were reduced throughout the 1990s as administrations cut education budgets as part of belt-tightening measures.

CONAIE's progress was gradual. President Jaime Roldós (1979–1981) established the Center for the Study of Indigenous Education at the Catholic University to address bilingual education and started a literacy project to train a generation of indigenous intellectuals. Roldós also created a national office of Indigenous Affairs in the Ministry of Social Welfare. (Indian issues had been dealt with earlier by the Ministry of Agriculture.) When CONAIE was founded in 1986, so too was the Scientific Institute of Indian Cultures (ICCI), which gave legal, ecological, and economic assistance to Indian communities. The ICCI was one of CONAIE's principal intellectual tools, and about 100 Indian students from various universities in Quito met regularly to assist in courses taught at the Institute.

Indian leaders were increasingly disappointed in the nation's leadership after the death of Roldós, although they were encouraged by some of President Rodrigo Borja's policies. President Osvaldo Hurtado (1981–1984), they thought, let his predecessor's programs languish, while León Febres Cordero (1984–1988) reversed many advances, suspended agrarian reform, and dealt harshly with land takeovers. President Borja gave CONAIE official recognition as the representative of the indigenous people, initiated a series of dialogues with its leaders, granted the Huaorani title to about half of their traditional territory, gave the movement the authority to name directors of bilingual and bicultural education programs, and at least entertained the idea of a plurinational state. All future governments would receive demands from CONAIE—not met until after 2000—for formal dialogue. Sixto Durán Ballén (1992–1996) was highly criticized by most indigenous leaders. He rejected the concept of a plurinational state and set up an office of his own to administer policies in the indigenous communities, naming José Quimbo of Otavalo as its director. CONAIE denounced the office and unsuccessfully sought to name its own representative to the government.

CONAIE grew stronger as the economy deteriorated in the 1990s and as government efforts to improve conditions in the countryside were diminished. Eduardo Gamarra, director of the Latin American Caribbean Center at Florida International University in Miami, claimed in 2000 that *Ecuador is a country that has basically forgotten about its agricultural and rural base, so conditions in the countryside have deteriorated in the last decade. The resulting process of pauperization in the countryside marginalizes indigenous people almost by definition, because of their strong links to the land.*[16] Many observers agreed. *The situation is truly tragic,* lamented Bishop Raúl López of Latacunga. *The government has cut the budget for everything in the social sector in order to satisfy the demands of the International Monetary Fund and to make payments on our foreign debt.*[17] Several Indians from Latacunga marched on Quito in January 2000 to complain that their villages were the last to receive electricity, water, telephones, and sewers.

Such was the social and economic setting as CONAIE stood poised for action at the end of the century. The movement, along with Pachakutic, the CSM, and other popular organizations, had largely replaced the traditional parties of the left and labor unions in expressing social discontent. The labor movement in Ecuador had always been one of the smallest and least influential in Latin America, owing to the country's late and relatively weak industrialization. Most analysts doubt that over one-fifth of the workforce was ever unionized. Some on the left claimed that CONAIE and other Indian organizations were merely concerned about ethnic revindication and the environment. Others disagreed; they saw the indigenous movement as going beyond ethnic interests and addressing national issues. In any event, the Indian movement was far more powerful than the traditional left had ever been. It had demonstrated and marched on Quito many times in the last fifteen years and moved the indigenous people from relative political isolation to the center stage. In the process it had broadened and deepened the imagined Ecuadorean community. The Levantamiento Indígena of January 2000 was to be its most momentous effort.

5

Bucaram, Arteaga, and Alarcón

Since the return to democracy in 1979, Ecuador's presidency alternated between parties popularly identified with the right and left of the political spectrum: Jaime Roldós and Osvaldo Hurtado on the left, 1979–1981; León Febres Cordero to the right, 1984–1988; Rodrigo Borja on the left, 1988–1992; and Sixto Durán Ballén to the right, 1992–1996. If the pattern held, in 1996 it was the left's turn. Oil continued as the economy's axis, and by the early 1990s the economic interests and social groups that the new wealth helped to create or transform had developed strong expectations of the government as well as an array of predictable demands based on self-interest. As a consequence, while petroleum prices declined for most of the decade, state spending and heavy borrowing from abroad continued unabated in a futile effort to meet those expectations and demands.

Nine men entered the May 20, 1996, presidential contest to succeed Durán Ballén, and from the outset it seemed unlikely that any of them would garner the 50 percent needed to win without a runoff. A contest between two coastal candidates, one conservative and one liberal, seemed likely to square off in the second round of voting. Jaime Nebot of the staunchly conservative Social Christian Party (PSC), who had lost to Durán Ballén in 1992, emerged as the early front-runner. Nebot had served as governor of Guayas province during the administration of his political mentor and sponsor, former president Febres Cordero, who was mayor of Guayaquil from 1992 to 2000. Nebot, a Guayaquil lawyer with strong business backing, vowed to create jobs and stimulate the economy even at the cost of higher inflation and a weaker currency. He advocated free trade and tariff and quota reduction, and firmly supported so-called privatization that favored private investors buying portions of state-owned enterprises, some in whole and others in part. Nebot considered virtually all state-run industries too costly for the government alone to manage, and he decried both their inefficiency and the inadequate capitalization that made it difficult for them to modernize their operations.

Nebot's strongest challenger was 44-year-old Abdalá Bucaram, who held a degree in law from the University of Guayaquil and another in physical education. His brother-in-law, President Roldós, named him Superintendent of Police in Guayaquil from 1979–80, and in 1981 he was made adviser to the country's inspector general. In 1984, Bucaram was elected mayor of Guayaquil at the same time that Febres Cordero appointed Nebot governor of Guayas province. He later served as a congressman. Bucaram formed the left-of-center Ecuadorean Roldista Party (PRE) in 1982, an organization named after Roldós, who was killed in a plane accident in 1981. Prior to the 1996 campaign, Bucaram had lost bids in 1988 and 1992 for the nation's top job.

Bucaram and Nebot were familiar names to Ecuadoreans. Both men were prominent descendants of Lebanese migrants who had arrived in Ecuador earlier in the century, a group that rapidly attained stature in both politics and business. At the end of the nineteenth century, when Turkey invaded Lebanon, a large number of Lebanese fled to Latin America and elsewhere. The Bucarams were perhaps the most successful of these Lebanese families.

Abdalá's cousin Averroes was president of congress, and many close relatives reached positions as deputies and ministers. Abdalá's sister Elsa and uncle Asaad both served as mayors of Guayaquil, where most of the Lebanese in Ecuador resided. Asaad, a national deputy and head of the powerful Concentration of Popular Forces (CFP) until his death in 1981, likely would have been elected president in 1972 and 1978 had the armed forces not blocked him, first with a coup and then with a regulation requiring that the parents of the commander of the Armed Forces (that is, the president) be born in Ecuador. The military's machinations to thwart Asaad's ambitions were encouraged by his chief rival in Guayaquil, Jaime Nebot Velasco, whose son now faced Asaad's nephew Abdalá in the 1996 election. In political terms, Bucaram and his extended family emerged from the CFP (established in 1950) with the support of marginal workers in Guayaquil. The party split into factions after 1979, and in 1982 Abdalá formed one of them, the PRE.

Political polls gave Bucaram the same 17 percent support he had won eight years earlier in the first tier of balloting. He was determined to win his third presidential bid. His stronghold was his native Guayaquil and the coast, where he outpolled all other contenders during the initial round of the 1988 elections. Bucaram's rhetoric against the oligarchy, his promises to lift those at the bottom of society, and his lewd humor all appealed to the poor, but polls showed he needed backing from more than the popular classes to put him in the presidential palace. To balance the ticket he turned to Rosalía Arteaga as his running mate. She was aligned with the political right, had an upper-class image, and attracted support in the more conservative highlands and from the women's movement.

Arteaga was a 39-year-old lawyer whose husband had been mayor of Cuenca, the country's third largest city after Guayaquil and Quito. In 1992 she took her first public post when President Durán Ballén named her assistant secretary of culture, and she subsequently served as assistant secretary and then as minister of education. She joined the conservative PSC in 1986, left it in 1991, and surprised most observers by accepting Bucaram's offer to run together in 1996. Identified with the upper echelon of society in dress and manner, she struck a conservative highland counterpoise to the flamboyant coastal candidate. During the campaign she was often treated in a bawdy way by Bucaram, who would laugh and hike up her skirt to let the crowds get a good look at her shapely legs. She put up with his antics without public protest. The team was, many commentators wrote, one of *la Bella y la Bestia*—Beauty and the Beast. Bucaram was backed by Álvaro Noboa, the richest man in Ecuador and heir to a banana fortune, and by Alfredo Adum, who reportedly contributed $4 million to his campaign.

Only months before the elections, Pachakutek decided to field a candidate. Since its founding in 1994 in the Amazon region, the party claimed to represent the indigenous population. The party backed Freddy Ehlers, the nominee of the New Country Party (NP) and host of a widely watched television show that addressed controversial political issues. Also in the running for the nation's top post was Quito's former mayor, Rodrigo Paz of the Popular Democracy Party (DP). Paz, who had made his fortune with money exchanges throughout the country, was a staunch advocate of privatization and was popular with the business community.

Bucaram and Ehlers competed for the same center-left vote. Bucaram, with the populist campaign slogan, *The Force of the Poor*, outpolled both Ehlers on the left and Paz on the right to qualify for the runoff round with conservative candidate Nebot. He did particularly well with poor farmers and urban slum dwellers with his attacks on the rich, promises of subsidies for items of basic consumption, and pledges to promote a variety of social services. As a candidate he vowed to use private capital for the country's development needs, to establish legal safeguards to prevent privatization from creating excessive private power and enrichment, and to represent the interests of the poor against what he referred to as the entrenched oligarchy.

Bucaram succeeded with unabashed rowdiness and populism. The highland elite was appalled by the prospect of a president whose nickname was "El Loco," but the poor of Guayaquil and other cities flocked to rallies where he sang, danced, joked, took the stage with a rock band and adolescents in hot pants, and occasionally pulled up his running mate's skirt. The two contenders were miles apart socially. As one national commentator put it, the boisterous Bucaram was not admitted to the private clubs frequented by Nebot, and his running mate later wrote that he often *ate with his hands* and *did not know how to use napkins.* As Bucaram phrased it, *I was unpredictable and definitely not one of them. I was not of their class, not of their group. The oligarchy has its ideas of equality. For them, the Indians have equality because they are all poor.*[1] In the runoff election on July 7, Bucaram picked up most of the votes that went to left-leaning Ehlers in the first round and beat the more conservative Nebot by a margin of 54.3 percent to 45.7 percent. His running mate, Ecuador's first female vice president, later wrote of herself that *Vice President Rosalía Arteaga in the campaign was fundamental, was crucial to the electoral triumph.*[2] Bucaram took office on August 10, 1996.

A BRIEF DANCE

August 10, 1996. President Abdalá Bucaram and Vice President Rosalía Arteaga celebrate after their inaugurations. The dance ended fifteen months later. Arteaga became Ecuador's first female president when she succeeded Bucaram, but she held the center of the floor for only three days.

The new president, like his predecessor, faced a badly fragmented legislature. Bucaram's PRE had 23 percent of the seats in the eighty-two-member Congress, while his archrival Nebot's PSC held 32 percent. The president's problem resulted from an inherent contradiction: his electoral support came from the center-left, but his economic policies, including privatization and modernization, drew legislative support from the right and opposition from the left. During his administration, coalitions were essential to pass most measures, and as always they were fragile and shortlived.

Latin American economies, according to the International Monetary Fund (IMF), were to grow at about 4.4 percent in 1997. Ecuador fell far short as growth reached only 2.6 percent in 1995, and a mere

1.9 percent the next year. The recession Bucaram inherited worsened as his administration unfolded. Oil prices slipped from $18.04 per barrel in 1996 to $15.58 in 1997. This price drop resulted from an excess of 800,000 barrels per day on the world market, occasioned in part by a 300,000 barrel-per-day reduction in demand. The year before Bucaram took office, the government ran its largest deficit in eight years. A projected inflation rate of 37 percent made access to credit difficult, as banks charged 48 percent interest on loans. With production and oil prices both in the doldrums, tax revenue plunged 19 percent in the first third of 1996, while the foreign debt rose 4.67 percent to $14.58 billion. The problems were long-standing. Bucaram believed they dated back to 1982, his brother-in-law Jaime Roldós's last year in office.

Although Bucaram campaigned on a populist platform, once in office he responded to the worsening economy by abandoning his plan for increased social spending. He pursued a conservative program based on balancing the budget, fixing the exchange rate to the U.S. dollar, and attracting foreign investment. The new president determined that the fundamental economic challenge was to increase confidence, production, and foreign investment, and to end financial speculation, inflation, and high interest rates. His predecessors had done much the same. In December 1996, Bucaram announced an austerity package, prepared largely by Domingo Cavallo, the former finance minister of Argentina, that included a convertible currency with a fixed exchange rate to the dollar and that made sharp cuts in state spending.

Nevertheless, Bucaram reaffirmed his pledge to help the less fortunate but told them to be patient. Significant improvements in their lives, he explained, would come in the long run, after fiscal discipline was instilled and the economy modernized. In the meantime, price increases were unfortunate but necessary to establish an economy without heavy subsidies and consequent price distortions. Many former supporters were outraged by what they considered Bucaram's about-face and false promises, and they accused him of buckling to the wornout failed attitudes and policies of the business elite. The president countered his critics by insisting that the economy and state had to be modernized and the nation's finances stabilized before Ecuador could advance and generate the new wealth needed to lift up everyone.[3]

By reducing expenditures to lower the deficit, Bucaram recognized that the inevitable price increases would provoke public discontent and would likely lead to labor strikes and antigovernment demonstrations.

Such were the invariable consequences of slashing subsidies for utilities, food, and other necessities. The cost of a tank of home cooking gas quickly soared from 2,900 to 10,000 sucres, even though the subsidized price was one-sixth of its production cost. Water and electricity rates jumped 115 percent; public transportation, 25 percent; and gasoline for vehicles, 20 percent. The concomitant increases in the cost of living, as predicted, sparked near-constant street protests, which began in earnest on January 8, 1997, and continued to build in momentum for the rest of the year. In 1996 polls had given Bucaram a 17 percent approval rating. In early 1997 the president's popularity rating had fallen to 12 percent. He was now losing ground even among his most ardent supporters.

Despite the diminished underwriting of various products and services, most outside observers thought that while the remaining subsidies were still substantial, further cuts would be difficult to make given the public's long-term reliance on the price-support system and its ability to effectively pressure the government by taking to the streets. They were correct. Bucaram claimed that underwriting gasoline was unfair, as it helped only 23 percent of the population, and that most subsidies went to those well above the poverty line. The same was true of other subsidies. The vast majority of consumers, however, focused only on rising prices.

The public thought that they were overtaxed, but analysts argued that they were not, at least in comparison to the majority of other Latin Americans. Ecuadoreans were accustomed to low taxes, owing in large part to the long reliance on oil and other export revenues, or to not paying taxes at all. In February 1997, Abelardo Pachano, former head of the country's Monetary Fund, lamented: *There are vast sectors of the economy with high revenues that have never paid taxes in Ecuador*.[4] Bucaram found out just how politically hazardous it was to increase them. His effort included higher duties on popular consumer items. The alcohol tax was upped 63 percent, the duty on light tobacco 103 percent and on dark tobacco 48 percent, and the levy for automobiles 10 percent. Declining revenue was not the result the government hoped for in curtailing subsidies and increasing taxes, yet state income plummeted 19 percent ($72.1 million) in the first trimester of 1997 compared to the same period a year earlier. During the first third of 1997, sales of alcohol dwindled downward to 60 percent. Claudio Patinio of the country's liquor association pointed out that the falling figures did not mean Ecuadoreans were drinking any less, but rather that sales of contraband liquor were soaring. Revenue Director Carlos Velasco Garcés called for the reduction of taxes on a number of items of popular consumption, insisting that lower rates would reduce contraband, increase the purchase of domestic products, and thereby restore lost revenue. On the left, critics of Bucaram's tax policies said he should have imposed taxes instead on the rich, on luxury items, and on tariffs.

Bucaram announced on December 1, 1996, a system of complete convertibility of sucres to dollars to take effect on July 1, 1997. All sucres in circulation would be backed by dollars, and new ones could not be issued without such support. The plan, along with tax and fiscal reforms, was to limit inflation and salary erosion, halt currency speculation, and encourage investment by closely linking the sucre to the more stable U.S. dollar. A *petroleum stabilization fund*, he argued, should be established to deal with the inevitable ups and downs in the economy and act as a financial cushion during times of falling export prices and diminished revenue. Convertibility, he claimed, the heart of his economic plan, would provide the country with economic discipline. Those who benefited from currency speculation, Bucaram argued, opposed his plan. Indeed, it was roundly condemned by a broad coalition that spanned both ends of the political spectrum and included most Chambers of Commerce and Industry as well as labor unions. The president approached Saudi Arabia for loans and claimed later that had he not been toppled, he likely could have obtained $600 million at 1.5 percent interest for fifteen to twenty years with a three-to-five-year grace period. (In contrast, the loans offered by the IMF, the World Bank, and other traditional lenders came with 9 to 18 percent interest and gave only one to two years of reprieve.[5])

During the 1970s, a portion of oil revenues had been invested to establish state-owned industries or to purchase shares in existing companies deemed essential to modernization. But beginning with Febres Cordero's administration, government participation in industry was gradually dropped in favor of a free-market approach to development, which advocates claimed would attract more capital and expertise, bring greater efficiency, and lighten the load on the already strapped budget. Ecuador still had about 200 fully state-owned firms and owned between 20 and 30 percent of a number of others. Bucaram and his advisers decided to privatize most of them, but exactly how to do it defied consensus. Debates raged over which enterprises should be sold and which kept, whether they should be auctioned off in whole or in part, what percentage of each should be retained by the government, how the value of the undertakings should be determined, what percentage of ownership should be allotted to employees, and whether nationals or foreigners should be permitted to purchase the enterprises.[6]

Ecuador tried three approaches. The first gave all ownership to the private sector. In the initial efforts, the government sold an airline and a cement business. The second approach was to encourage private capital in new projects while maintaining state ownership of similar ones already in existence. President Febres Cordero had acted along those lines to obtain new investments in petroleum, electricity, and telecommunications. The third was to permit private investment in existing enterprises, converting them into mixed ventures. Telephone and electricity companies were the ones most often mentioned for this tactic.

The efforts of Bucaram's predecessor, Durán Ballén, to diminish the state's direct role in the economy had met with opposition from practically every interest group, including the Armed Forces, trade unions, and indigenous organizations. When the military ruled from 1972 to 1979, the generals did not hesitate to use a portion of the new oil revenue to advance their interests, and they purchased shares in defense industries, an airline company, banks, and even some hotels and restaurants. They opposed Durán Ballén's privatization initiatives as detrimental to their own privileges and income. For their part, labor unions and Indian groups thought privatization would result in higher prices and reduce the government's ability to redistribute income. These opponents believed that privatized companies would no longer serve the state, provide jobs, and offer low-cost products but instead would answer only to investors whose sole concern was profits.

Opposition parties in Congress managed to block large-scale privatization proposals as public opinion polls still revealed heavy opposition to the general notion together with divergent views on how it should be implemented. The fear that vital sectors of the economy might fall into foreign hands was another underlying reason for the formidable opposition. The net effect of Durán Ballén's efforts was that instead of a projected 80 percent of state companies passing into private hands, he presided over the transfer of a mere handful. The bankrupt national airline was the only major one relinquished, although some smaller interests in hotels, restaurants, and road-building companies were sold off as well.

Bucaram's critics charged that he broke his campaign promise to use the country's existing private capital for development and that he failed to establish effective legal safeguards for privatization. Ecuador's inefficient state-owned telecommunications industry (EMETEL), for example, was one the president proposed partially to auction off. He hoped to sell 35 percent of its shares to a private foreign bidder; of the remainder, 10 percent would be distributed among its workers, and the state would retain a 55 percent stake. Opponents attacked his plan from all angles, wanting assurances that efficiency, lower prices, or other benefits would accrue. Bucaram's response was that only his approach could gain international confidence, secure private investments, and win additional foreign loans. He dismissed attacks against him and his programs as nothing more than unscrupulous allegations of political aspirants who had suffered defeat in the last election. Privatization, he argued, was essential for efficiency and modernization, and it would also provide, he hoped, much-needed government income for social development. Nevertheless, he failed to sell off the telecommunications industry.

It seemed to some observers that the executive had made an effort to alienate not only the elite but also virtually everyone else. He attacked the press, mocked opposition parties, and condemned the so-called oligarchy. He angered nationalists by strongly urging that Peru and Ecuador forgive and forget their past differences. The flamboyant leader even invited Lorena Bobbitt for lunch at Carondelet. The famous Ecuadorean had achieved a measure of international notoriety by cutting off her American husband's penis with a kitchen knife.

Of greater political importance were rumors of corruption concerning Bucaram and his extended family, which drew almost daily news coverage. Seven years earlier, Bucaram had gone to Panama after losing the 1988 elections and being accused of misusing public funds. His sister Elsa was still living in Panama, where she fled before resigning as mayor of Guayaquil in 1991 when faced with charges of stealing millions of dollars. And the president's 19-year-old son, Jacobo, was rumored to have garnered millions of dollars from a post in customs. The allegations of corruption against his family did enormous damage to Bucaram by undermining his moral authority and capacity to lead.

The public had become accustomed to reports of high-level corruption as well as to the standard retort that the allegations were politically motivated. Former President Hurtado said that corruption had become a cultural problem that permeated the entire society and not the public sector alone. He believed that it followed centuries of poverty and low government salaries. Others lamented that corruption and cynicism had so penetrated the business community that honesty was equated with being a poor businessman. (In mid-September 2000, a survey of Transparency International, a private institute based in Berlin, would combine the perceptions of business executives and analysts with public opinion polls concerning ninety countries. The survey named Ecuador the most corrupt nation in Latin America and Chile the least so.)

After leaving office, Vice President Arteaga would write that during Bucaram's tenure *the 10 percent commission on all contracts . . . had risen to levels of 15, 20, and up to 30 percent*,[7] and many observers deemed theft during Bucaram's administration more widespread and serious than usual. Whether it was or not, by January 21, 1997, U.S. Ambassador Leslie Alexander and the chief of the Southern Command of U.S. Forces, General Wesley Clark, were convinced that the practice was completely out of hand. Alexander and Clark met with General Paco Moncayo, head of the Joint Command of the Armed Forces, along with other top military leaders, for the sole purpose of discussing the rampant stealing. Moreover, Alexander, in a public address on January 29, spoke out, saying that government corruption was hampering the implementation of economic measures. *Ecuador is gaining a reputation for pervasive corruption*, lamented the ambassador. *A $12,000 bribe was demanded of a businessman to get an $8,000 container out of customs. News of this sort of mad extortion, which defies even the usual sordid conventions of corruption, reaches international corporations very quickly.*[8]

Four days later, the nation's top uniformed officials presented Bucaram with information about major improprieties at the customs house in Guayaquil. Their charges included the names of corrupt officials as well as the amounts of money illegally taken, and they insisted that the head of state explain the irregularities. Bucaram's response was a promise to look into it. Nothing changed. Mismanagement at the customs house gained heightened media attention. On the highly charged issue of corruption, Bucaram pointed out that it had always been an endemic national problem. One must recognize, he stated matter-of-factly, that *everyone robbed*. Corruption was an age-old and worldwide problem that was understandably acute in poverty-stricken developing countries.

In foreign policy, the President repeatedly urged peace with Peru, calling the border conflict *stupid wars that have no reason for being*. He lamented and discouraged an arms race between the two countries: *Ecuador can never advance, nor can Peru, while their Armed Forces continue on a war footing. One cannot seek peace by buying arms and trying every day to become stronger to defend the frontier*. Defense spending stood at around $720 million. The 300,000 houses that the president proposed to build for the poor would cost roughly the same amount as the military wanted for new equipment in 1997. He thought that Moncayo was obsessed with military purchases and complained that

GENERAL PACO MANCAYO

A war hero and head of the Armed Forces in 1995, the general became a power broker in 1997 when three people simultaneously claimed the presidency. He was elected to Congress in 1998. Expelled in 2000 after submitting his resignation to the Junta of National Salvation on January 21, he was elected as mayor of Quito in 2000.

at least twice per week the general would tell him how far ahead Peru was in one armed capacity or another. Yet Bucaram refused to engage in an arms race, saying it was impossible *in circumstances under which there was not enough money to pay for public works nor to pay government salaries*, and insisted that diplomacy was the best course.[9]

From the outset of the conflict with Peru, the Indian movement firmly opposed war and called for a cease-fire. CONAIE was pacifist in orientation, and over 300 Indian communities were located on both sides of the border in the area of likely military confrontations. Luis Macas wrote: *We are Indigenous Peoples to which no government, neither Peru's nor Ecuador's has paid attention.*[10] Furthermore, at the time Bucaram was inaugurated, members of the military had not been paid for three months, although they had received their salaries promptly during Durán Ballén's administration. Bucaram believed withholding pay was designed to undermine his relations with the Armed Forces from the outset, thereby destabilizing his government. On January 14, 1997, a mere month before the general strike, Bucaram traveled to Peru to address its congress and call for peace and mutual forgiveness. While he was well received in Lima, his enemies at home *asked for my head for speaking of forgiveness between Ecuador and Peru.*[11]

Foes in Congress accused Bucaram of unprecedented corruption and of worsening the country's economic ills through his erratic behavior and political showboating. Protests across the country came with increased frequency. One demonstration followed another to denounce nepotism, corruption, and austerity measures that meant higher taxes, fewer services, and less purchasing power, all in contradiction of Bucaram's campaign promises. The president did not resort to repression to quell the discontent for fear such action would result in his overthrow.

On January 25 top military commanders met with Bucaram to inform him that antigovernment sentiment was growing significantly deeper and more widespread, and that hostility to the administration pervaded the country on all social and economic levels. They reported that the opposition had turned to genuine public anger and was growing larger and stronger daily. The revulsion, they revealed, was aimed not only at the regime's economic measures but also at the corruption, nepotism, and arrogance of the executive and his closest companions.

The president believed his two major problems were the public disenchantment with his economic policies and the political stalemate in Congress. He saw the latter one as institutional in nature and urged a constituent assembly to restructure the political system. Convinced his economic austerity measures were essential, Bucaram understood there would be protests about them. He had familiarized himself with the recent events in Mexico and Argentina—two countries that also saw demonstrators take to the streets against unpopular but necessary spending cuts. Bucaram recalled that when President Hurtado eliminated gasoline and wheat subsidies in the fall of 1982, in the wake of a national shutdown called by opponents of the measures, a national emergency and five days of curfew followed. There had been, he pointed out, at least twenty nationwide general strikes since Hurtado left office in August 1984. The upcoming one, he believed, would be just another in a long line. He was, after all (with the exception of Jaime Roldós, who was killed in a plane crash), the sixth democratically elected leader since the military relinquished power in 1979.

The president developed a strategy designed to handle either a weak or a strong strike. If the strike fizzled, he would likely not change economic course. If it proved powerful, he would grant wage increases to dissipate or end the shutdown. He believed that if forced to, he could unite his government with labor and survive the challenge. Far from criticizing the workers' movement, Bucaram lauded its efforts as a sign of political maturity and declared that along with other disadvantaged groups, labor had learned to defend its interests.

On February 5, 1997, the near-unanimous opposition to Bucaram came together to paralyze the country with a general strike, called by the United Workers Front (FUT), scheduled to last forty-eight hours. Coordinated by an umbrella organization called the Patriotic Front, it was backed by business and labor alike. The Indian movement CONAIE, the Coordinated Social Movements (CSM), the Women's Political Movement (CPM), and a host of other entities joined the shutdown. Pressure on powerful chambers of commerce and industry to take part was applied by Guayaquil's mayor, Febres Cordero, who was decisive in gaining their cooperation. The former president and archenemy of Bucaram had said after his foe's election that *only prostitutes, marijuana smokers, and vile thieves* voted for the new president.[12] Febres Cordero and two other former heads of state and political rivals, Hurtado and Borja, now joined in asking Bucaram to step down.

On February 2, Vice President Arteaga had released a statement to the press critical of the administration. She did so despite urging from top military and Church officials that she exercise *prudence* and *patience*, and her belief that *because there was an explosive climate in the country, whatever I said could be magnified or be interpreted in different ways*. She feared that *given the unstable personality of Bucaram, nobody knew how he would react* to her efforts to *separate myself even more from what the government was doing*. Arteaga insisted she *should have been consulted on all government decisions, but I never was.* She was concerned about corruption; disagreed with Bucaram's economic measures, price increases, and plan to make the currency convertible; and thought that his having asked for pardon from Peru had been a national humiliation. The vice president claimed she had not criticized the regime earlier because *I always had a preoccupation: the preoccupation that they were going to see me as ambitious. . . . When they speak of masculine ambition it appears to be a virtue; on the other hand, the ambition in the case of Rosalía Arteaga was always seen as a defect.* Arteaga nonetheless recognized that for many people, *Bucaram and Arteaga function as a pair.* When she asked Quito's Mayor Jamil Mahuad to lunch in late 1996, for example, he refused. *I believe he was afraid of being identified with me,* she recalled, *as I had evidently fallen in disgrace along with the regime.* On the other hand, she thought the labor movement viewed her favorably, that it felt *solidarity with respect to a person not contaminated with the government of Bucaram.*[13]

When the strike was launched on February 5, the wide range of interests united against the president had hardened and radicalized their positions. They were in no mood to compromise; they were determined to drive him from office. Two million people participated in the two-day strike. *No One Stays Home* was its slogan, and leaflets circulating before the event declared: *We invite all Ecuadoreans to the giant going-away party for Bucaram and his family on their one-way trip to Panama or wherever. This event will take place in the country's plazas and streets on February 5 and 6. Dress informally. The entrance fee is a street barricade, a burning tire, and the will to save the country's dignity.*[14]

In the capital and other major cities, CONAIE worked closely with the CSM to close banks, factories, shops, financial markets, and industries and to paralyze public transportation. In addition to the urban demonstrations, CONAIE blocked the Pan American Highway connecting provinces to the north and south of Quito. The situation grew increasingly tense when Bucaram decreed a state of mobilization, which allowed him to use troops to maintain order.

The antigovernment movement turned out to be bigger and broader than anyone had predicted. To Vice President Arteaga, it reflected the *collective nausea* of all social strata against Bucaram's policies and personal conduct. Although she refrained from expressing her view in public, she privately *believed his departure was absolutely necessary for the country.*[15] Pichincha Chamber of Industries leader Gustavo Pinto remarked that he never imagined the strike would reach such enormity, and Quito Chamber of Commerce head Nicolás Espinoza claimed that only afterward did the business community recognize its size and significance. Some television stations announced that one million people had taken to the streets, while others doubled that figure. Mayor Mahuad led tens of thousands of people in the capital who flooded Avenida 10 de Agosto, long known as the "political street of Quito." Industrialists and other business leaders refused to aid Bucaram, who had labeled them oligarchs and worse.

Bucaram addressed the subject of the strike in a televised address to the nation on February 5. *The national government respects the positive aspects of the peaceful movement of February 5,* he said. *We understand that this strike is a historic protest against backwardness, underdevelopment, degradation, and the decadence of the political regime, and the domination and exploitation of the people and the nation.* In the same speech, the only solutions he offered were modernization and more foreign and domestic private investment; no mention was made of modifying his unpopular economic measures or making cabinet changes. Bucaram put the number of demonstrators at around 150,000, not one or two million, and contended that most of the country was tranquil. He nonetheless admitted that the strike was a success in Quito and Cuenca. *When the strike ended,* he continued, *I knew that only a miracle would impede the fall of my government. Moncayo no longer obeyed my orders, and Febres Cordero and Alarcón had arrived at an understanding. I accept that many people who voted for me thought that my government was not responding to their expectations; I also accept that among them there were those who were in agreement with my being deposed.*[16]

As tens of thousands thronged the streets of the capital to demand that their leader step down, many looked to the Armed Forces to resolve what they were sure was a mounting crisis of major proportions. Troops stood poised in the barracks and ready for action with a security strategy to maintain order. The plan was not needed. Like the rest of the country, the generals and admirals were unanimous in their opposition to the executive, and the unexpected magnitude of the strike confirmed their growing belief that Bucaram could not govern effectively. The solution to the impending chaos, they calculated, was to withdraw their support from the president, hope he would resign or ease him out by other means if he did not go willingly, and let professional politicians pick his successor. Removing their backing was easy; the rest required rigor.

As the shutdown grew near, the president confided to his closest associates that he would likely announce salary increases and give union leaders credit for the concessions. He was confident the action would satisfy the workers and strengthen his own position as well. The only condition he planned to impose in return would be that labor leaders condemn the presence of *political opportunists* in the strike. The president's delusion that the difficult situation could be controlled continued until the evening of February 6.

During the administration's first few months, Congress had considered but did not adopt a proposal introduced by Deputy Marco Proaño Maya to enable the legislators to select the president in the event the office fell vacant. Vice President Arteaga condemned the effort in a conversation with Bucaram, who told her he too opposed the plan. She was nonetheless convinced the president was behind the effort, which he was. The notion that Congress should name a new executive was revived at least a month before the February shutdown, as deputies searched for a way to remove Bucaram. The major impediment to Bucaram's ouster was that the constitution required a two-thirds vote for impeachment of the president by Congress for treason, bribery, or *any other infraction that gravely affects the national honor*. As lawmakers cast about for a constitutional basis for removal, it was clear that proving impeachment charges would entail a lengthy trial and that, more to the point, they lacked the fifty-five votes required to effect it.

Article 100 of the national charter read that a simple majority could remove a president on *account of physical or mental incapacity declared by the national Congress*. The country's leader had for years been dubbed "El Loco" owing to his unconventional style and lack of inhibition, and the deputies who embraced the article had three more votes then needed to remove him. Although the constitution did not elaborate on mental incapacity, as Bucaram later pointed out, the nation's civil code provided for qualified professionals to conduct an examination and make a recommendation to a judge. The code allowed an appeal of the judge's decision. No such process was followed as Congress made its purely political decision. Bucaram claimed that Ambassador Alexander and General Moncayo both had told him that Article 100 did not apply to him. Vice President Arteaga later wrote: *The Armed Forces were aware that this was not the correct way of removing Bucaram, that it was being done due to social and public pressure that grew daily. The Armed Forces could not find a solution either*.[17]

With the problem of how to depose the president apparently solved, the next challenge was selecting his replacement. The deputies coordinating the executive's removal floated the idea of giving the vice president power for up to 180 days, during which time elections would be held. The proposal was rapidly dropped as it became apparent that Arteaga did not have the support of the majority of deputies. Many feared that if she were allowed to take over, there would be no way to get her out before the end of Bucaram's unfinished term in 2000. On February 5 the vice president reacted to reports that Congress intended to pick the president's replacement by calling it an *unconstitutional coup d'état*.[18] She later said Defense Minister General Víctor Manuel Bayas and General Moncayo had urged her to condemn the congressional effort to name Fabián Alarcón president; they agreed with her that while Congress could depose a president, it was without constitutional authority to elect one. The next day, February 6, Arteaga appeared on television and repeated her position. She saw Alarcón as closely connected with Bucaram's regime, as Bucaram had backed Alarcón's election as president of Congress three times, had *never pronounced against Bucaram*, and *during the six months of Bucaram's government they had formed a harmonious and compatible duo*.[19] Bucaram had given Alarcón a major voice in naming appointments to the national postal services and several posts in EMETEL and other institutions.

Deputy Álvaro Pérez had proposed earlier that Congress name Alarcón president by simple majority vote. Although Alarcón was the head of the legislative body, not everyone backed him, as he lacked a popular following and his small conservative Radical Alfarista Front (FRA) claimed only two deputies. Alarcón consented to the plan to vote Bucaram out and himself in, but only with signatures on a document showing a congressional majority for the proposal. The signed pledges were not obtained until February 3.

Fully aware of the congressional maneuvering against him, on February 4 the president assembled his top aides Alfredo Abdum Ziade, Miguel Salem Kronfle, and Eduardo Azar. He also met with a number of PRE deputies and other backers to devise a strategy to ensure enough votes to block removal on the grounds of *mental incapacity*. He decided to implement the contingency plan made earlier, to accommodate the strikers to defuse the opposition. In the process, he would derogate the most criticized of his austerity measures and hope for the best.

Bucaram invited television reporters and their cameras for his February 6 announcement that the highly unpopular tax on electricity would be reduced 25 percent and that the even more despised gasoline tax would also be lowered. He later said his gasoline price hike *was an error, the product of bad information. . . . I realized it too late, after the economic package was launched*.[20] The president also declared it was imperative to raise salaries 25 percent, and he suggested that his administration was considering abandoning all of its austerity measures and starting anew. Two unpopular ministers would resign: Abdum Ziade and Salem Kronfle.

On the same day that Bucaram pledged to cut taxes and Salem Kronfle persevered in fruitless efforts to isolate the PSC, a momentous vote was taken by the deputies, with Raúl Baca temporarily entrusted with the presidency of Congress. The Armed Forces had already given up on the executive and had conveyed its position to legislative leaders. The vote came at a little past 10 P.M. on February 6. It removed the colorful leader on the basis of mental incapacity, made Alarcón interim president, and called for elections in August 1998. Forty-four voted in favor, thirty-four against, and two abstained. In short, the deputies had three more than the simple majority of eighty-two required by the constitution. Of the forty-four votes, twenty-four came from the PSC, eight from DP, six from Pachakutik, and two each from the Democratic Left (ID), the FRA, and the Popular Democratic Movement (MPD). Bucaram later claimed many deputies sold their votes: that one told an intermediary that while the PSC offered to pay $500,000 for his vote, he would support the president for $750,000. Vice President Arteaga subsequently wrote that Bucaram had told her many deputies were charging over $1 million to vote in his favor. The allegations were both denied and unsubstantiated.

At 10:04 in the evening the president of Congress thus became president of the republic—or so he and most deputies believed. The motion read:

> Abdalá Bucaram Ortíz constantly and repeatedly violated the constitution and the laws of the republic, trampled civil society, menaced and assaulted the press and other communications, utilized the Armed Forces for acts foreign to their functions, prejudicing their image, acted in artistic, musical, and sports events at odds with the dignity of the presidency of our republic, and has set up a giant web of relatives and friends who are the center of corruption which, converted into a system of government, flagellates and denigrates Ecuador. . . . The virtually unanimous urging of the population is to terminate this state of chaos and outrage organized by the executive.

It continued: Congress *renounces the convertibility project*. Moreover, it demanded a *dignified* and *definitive peace* with Peru that *recognized a sovereign territorial outlet to the Amazon*.

In the meantime, the military redoubled its contingent around the legislature, and Alarcón as well as the parliament received assurances of their personal security from the high command. The head of the FUT informed the press that General Moncayo had told him *the people themselves have made him no longer legitimate*.[21] Ambassador Alexander had also been speaking with the general, informing him that it would be difficult for the United States to maintain relations with anyone who took power in such dubious legal fashion as had Alarcón. The changing of the guard was not as smooth as the civilian and military planners had hoped.

Ecuador found itself on February 6 with three individuals who claimed the country's top office: Bucaram, who had not resigned; Alarcón, named by Congress to replace him; and Vice President Arteaga, who argued her office put her next in line for the job. Arteaga said she announced she was president when General Moncayo told her *the high command would not take my name in consideration unless I put it into play*. She also said that Alexander phoned her and referred to her as *Señora presidenta*.[22] Each claimant offered a radically different interpretation of the law. While Alarcón accepted Congress's actions, Bucaram and Arteaga adamantly dissented.

Bucaram claimed the congressional votes putting in Alarcón were contrary to the constitution and that therefore he was legally still in command. He refused to relinquish his office and vowed to finish his term. Arteaga did not protest the president's removal but proclaimed herself the nation's head because she had been elected by all the people; further, according to the constitution and by precedent, vice presidents succeed presidents when vacancies occur. Privately, Arteaga *believed Bucaram's departure was absolutely necessary for the country*, but she did not publicly urge him to resign.[23] She remained in her office, ready to enter Bucaram's. Arteaga and Bucaram had been publicly estranged since September. As his running mate she had called him moderate and measured, but now she confessed she had been fooled. As for Alarcón, he simply said Congress had elected him, but nonetheless he recognized that the governing charter did not give the deputies power to elect the president. The legality of the recent events was becoming murky as raw political power and aversion to Bucaram trumped constitutional procedures. That much was growing clear to soldiers and civilians alike.

Thousands thronged the capital's streets on February 6, and a large crowd formed at San Francisco Plaza and struggled past security forces to advance a few blocks to nearby Independence Plaza in front of the presidential palace. As the crowd swelled, General Moncayo doubled the police and military force around Carondelet. He phoned Alarcón in Congress to chastise him for making statements about who was in charge and who had to go, which Moncayo said incited the impatient populace. Those who gathered listened to speeches that applauded

Congress's actions and demanded that Bucaram accept his expulsion. Certain that violence would erupt, the president asked the military to use tear gas to keep order. The crowd dispersed and barbed wire was put in place to keep demonstrators away.

The Armed Forces, fearful that public demonstrations would get out of hand, warned of the chaotic consequences of a struggle for power and urged Bucaram, Arteaga, and Alarcón to find a speedy solution to the crisis. At 3 AM in the morning on February 7, a few hours after Congress voted Alarcón president, General Moncayo issued a statement saying the Armed Forces did not recognize any of the three contenders and called for a *national dialogue*. Arteaga thought *the situation was turning unsustainable from both the internal and international points of view*.[24]

Despite fears the military would intervene to resolve the presidential dispute, Moncayo said his forces would remain neutral; so too did the head of the police, General Marco Hinojosa. The U.S. government, however, did not stay on the sidelines and weighed in heavily in support of what it called the constitutional process, which it claimed was being violated. Ambassador Alexander informed Moncayo that it would be difficult to recognize Alarcón's regime, as it had not been formed pursuant to the nation's governing document. After talking to Alexander, Moncayo and the top military echelon became even more preoccupied with the international repercussions of how Bucaram's successor was selected. Earlier in the day on February 6, members of the Joint Command asked the heads of the country's various military districts for a sense of the popular mood. All reported their districts strongly hostile to Bucaram. With their evaluations confirmed, the generals and admirals met in the Defense Ministry to determine their response to the three-way power struggle.

On February 7, Bucaram called Arteaga to request an urgent meeting. In the vice president's office, the president asked her for her support in return for control of seven government ministries. Bucaram claimed they could govern together in harmony and boasted that he would win a new vote in Congress because money from banana magnate Álvaro Noboa would buy enough million-dollar deputies to back him. Arteaga's answer was that time had run out, and Bucaram angrily ended the half-hour-long conversation.

Later that day, General Moncayo and fellow Joint Command generals Hernán Quiroz, Cristóbal Martínez, and César Durán Abad informed Bucaram that he no longer had their support. The executive's response was to make three points and a plea. He would not resign; he had been elected directly by the people, and Congress could not undo that vote. His unpopular economic decisions were necessary, he said, and he asked the officers for ten days to communicate with the country and to return it to normal. He assured them he would succeed and later wrote that he planned to convoke a meeting of labor leaders, ethnic groups, and Chambers of Commerce and Industry to reach agreements on solutions to the nation's problems. The commanders reiterated that the Armed Forces had made its decision, however, and gave the president no indication that his departure would be reconsidered. Within hours the president's wife was packing suitcases in the presidential palace.

That evening General Moncayo called Arteaga at 6 PM: *Bucaram is going to hand over power to you, and with that he will leave Quito and we will have secured tranquility in the country*. But when the vice president, the high command, and the president met shortly thereafter, Bucaram announced he would step down with conditions that no one in the room expected. According to Arteaga, he said: *The little woman has obtained what she wants, I am going to give her power. But I will be in Guayaquil watching that she does not change a single minister, only perhaps those I have decided to change myself; we will not modify the economic plan, there will be no change in authorities in Guayaquil, and I will speak from Guayaquil when I want. Power will be given to her every fifteen days*. The agreement to turn over power would be verbal; nothing was to be put in writing. Arteaga claims General César Durán advised her, *You are not obliged to accept, we are really surprised by what happened*; and that General Moncayo chastised Bucaram by saying, *Abdalá, that is not what we agreed on, the situation is totally unacceptable; you cannot think of verbal entrustment, you cannot think of entrusting in that manner*. Arteaga believed that *Paco Moncayo did not know what to do*. Indeed, *Bucaram fell victim to his internal passions, and at some moments he felt the best way out was to cede to me; in others he felt it was his end . . . he was in a contradictory state*.[25] The meeting ended without resolving who would be president.

Bucaram left Quito at 7 PM, and the presidential palace was immediately put under military control. He flew to his hometown and political stronghold, Guayaquil. At the airport in Guayaquil, only seven soldiers stood ready to offer him security in further evidence of the military's decision to abandon him. Perhaps in anticipation, Bucaram had his own bodyguards with him. He told reporters he was in the port city for the pre-Lenten Carnival festivities that began that evening, but he issued an ultimatum to Alarcón to meet him in Guayaquil the next day. In the combative style marking his political career, he lambasted Alarcón for usurping presidential authority. *Only I am the president of this republic,* he raged; *if Alarcón were a man* he would not be afraid to meet face-to-face to either settle things or give up his aspirations. According to Vice President Arteaga, *Everyone believed Bucaram was going to make himself strong in Guayaquil, that we were going to have a civil war*. In a measured response to Bucaram, Alarcón tried to draw a contrast between "El Loco" and himself in rejecting the challenge. *The people don't want to see two bad boxers in a street fight,* he answered, *but statesmen who know how to resolve the country's problems.*[26] Despite his bravado, the embattled executive knew he was defeated and searched for a way to negotiate his departure under the best terms possible. General Moncayo spoke to Bucaram's top aide, Alfredo Adum: *For God's sake, make him understand that he has to leave because the people are in the streets and do not want him to continue as president.*[27] Although he refused to relent, Bucaram quietly made financial and other preparations to leave Guayaquil for Panama City.

The February 8 headline in Quito's newspaper, *Diario Hoy*, reflected both the powerful role of the Armed Forces and the enormous prestige of General Moncayo. *Where is the hero?* it asked. The name came from the jungle area of Cenepa on the southern border where Peru and Ecuador had fought two years earlier, and where Moncayo and his troops had successfully resisted Peruvian forces. There had been unconfirmed rumors during 1996 that the general, "The Hero of Cenepa," was considering a coup owing to his disgust with Ecuador's political leaders. *Diario Hoy* now criticized the Armed Forces for not exercising its *historic role* and forcing Bucaram to turn over the presidency to Alarcón. The story concluded: *The eyes of the whole country are turned to the chief of the Joint Command of the Armed Forces.*[28] Bucaram later wrote that he believed General Moncayo wanted the presidency himself but instead was thrust into the role of arbitrating a struggle between Alarcón and Arteaga over who would take command.[29]

In fact, the general was acting with dispatch. On February 8, Defense Minister Víctor Manuel Bayas left his post and General Moncayo temporarily assumed it until the new president named a successor. Moncayo was positioning himself to dictate a settlement between the three claimants and tersely announced that the country could not have a power vacuum, while stressing the military's neutrality in the dispute.

Moncayo talked frequently to Bucaram, Arteaga, Alarcón, Ambassador Alexander, and others and drew four conclusions. First, the pressing political problem lay not with Bucaram's ouster, as only he and his supporters challenged the constitutionality of his removal. Second, the Armed Forces could not support Alarcón under the conditions in which he had been chosen without taking unacceptable international risks. For example, the U.S. government threatened to withhold diplomatic recognition, which would make peace with Peru exceedingly difficult to achieve. Much-needed foreign loans and investments, moreover, would likely be impossible to obtain. Ecuador, the generals judged, simply could not withstand the threatened diplomatic and financial isolation. Third, although the constitution was ambiguous on the issue of presidential succession, there was a long-standing tradition of vice presidents taking over when the top position fell vacant. That order of succession was the emphatic preference of the U.S. ambassador. Fourth, the majority of congressmen, the heads of the most powerful political parties, former presidents Hurtado, Borja, and Febres Cordero, Quito's Mayor Mahuad, and the nation's powerful Chambers of Commerce and Industry all backed Alarcón. Arteaga later reflected on her position at the time: *Once again, I was alone before all the political parties, before the entire political class . . . in Congress, all the deputies continued speaking in favor of Alarcón and against me. . . . I can imagine the enormous pressures that weighed on the Joint Command.*

Arteaga was an unacceptable choice to many important interests. Her reputation and popularity with most deputies was low, and support was not much higher with the business community. The consensus was as a part of Bucaram's inept and corrupt government, she could not lead the nation. During the street demonstrations in Arteaga's home town of Cuenca, many people had shouted: *Bucaram has no value, neither does Arteaga.* Moreover, a majority of the public did not believe she had the personal or political capacity to serve. Others, including Arteaga, believed blatant bigotry was at work and that she was unacceptable simply because she was female. The problem with the solution concocted by the general and his colleagues was that Arteaga wanted the office to which she felt legally entitled. While the two other claimants hurled insults at each other, the vice president worked quietly behind the scenes, talking with such powerful figures as the Ambassador Alexander and General Moncayo. The U.S. officials equated her ascension to the presidency with the constitutional process, a fact she repeatedly conveyed to the general and others.

Moncayo and the uniformed officials decided Congress's authority to depose Bucaram would be recognized and the vice president would be sworn in. Arteaga claims that on February 8, Moncayo phoned her to say that Congress was looking for a solution to the crisis: *You should not negotiate, the Joint Command will speak with the deputies; do not negotiate, the deputies want to win quotas from you; do not negotiate with them because we know how we are going to do it; it is already in the hands of the Armed Forces*. Later, at 3 PM, Moncayo informed Arteaga that Congress would give her power. When she asked him how that could happen since the majority were against her, Moncayo answered, *No, leave it to us, . . . we have spoken with certain sectors and they are going to cede power because it is the only legal way out*. To her question, *What is going to happen afterwards?* Arteaga relates that the general replied: *Congress will entrust you with power; later they will meet and modify the constitution and that modification of the constitution can be made in the time they believe is convenient, and after that modification Congress will elect the president. And they will elect who they want; if it is Alarcón or another person, that is not important to us, but there has to be a legal way out*. She should, she was told, negotiate only with Alarcón. Arteaga claimed, *I noted that afternoon his attitude* [Moncayo's] *had changed, and in truth I cannot explain exactly why*.

It seemed simple, provided Arteaga went along. She believed Moncayo had promised she would retain power for as long as it took the deputies to revise the constitution to empower the legislature to elect the president. Under that condition she consented to the arrangement. Alarcón and Arteaga met for two hours on February 8, and the vice president emphasized to Alarcón that she and Moncayo agreed his election was unconstitutional. Arteaga claims that Alarcón *proposed an enormous government quota for me if I accepted him as president*, suggested that she remain as vice president and that he be interim president, and that a national referendum be held on the single question of whether the nation accepted his presidency and Bucaram's ouster. Arteaga countered that the constitution did not contain a provision for an interim president and that any plebiscite should offer voters the choice of recognizing her as president. Alarcón refused, and Arteaga then told him *that it would be important for him to speak with the Armed Forces who in the end would decide*. At about 9 PM that evening, she relates that Moncayo phoned to say: *I congratulate you as president, you are going to be designated; the deputies have returned to Congress, and you are going to assume command*.

General Moncayo, Alarcón, congressional leaders, and Arteaga finally agreed to elevate the vice president. Moncayo assured the other notables that if Arteaga were sworn in, she would step down soon. The general made the pledge to fellow military chiefs as well as to Alarcón, congressional leaders, and others including Quito's Mayor Mahuad. After her resignation, the original choice of Congress could rapidly and legally be elected interim leader, with elections to follow within a year. That same day Congress met. After Alarcón resigned the interim presidency, the deputies voted in Arteaga as interim president.

We waited for a communication from Congress that would indicate to us precisely the decision of Congress in entrusting me with power, Arteaga writes. *The document did not arrive . . . which demonstrated a lack of respect from Congress.* Arteaga nonetheless immediately made plans for a swearing-in ceremony the next day, under the clear impression that *the agreement I had arrived at with the Armed Forces was to remain in power until the constitution was reformed and a president was constitutionally named.* When the resolution passed by Congress still had not arrived early on the morning of the 9th, she *had the sensation that I was being fooled.* The document arrived shortly before she was sworn in at 10:30 AM, and ratified Bucaram's ouster on the grounds of mental incapacity. It gave *the presidency of the republic, in temporary form, to Doctor Rosalía Arteaga Serrano, constitutional vice president of the republic, for a limited time strictly necessary and indispensable, until the national Congress designates a constitutional interim president of the republic.* Arteaga would resume the vice presidency once Congress selected an interim president.

Arteaga claims that Moncayo told her, *What is happening is that you are terrifying them because you give the impression that you are preparing to remain in power a long time*, to which she replied, *It is not important that it be one or two days or longer and I believe that it is going to take them more time to modify the constitution.* Moncayo retorted, *This is only going to be two days . . . they will not reform the constitution.* Arteaga was stunned and could only respond, *That cannot be.* She later admitted to being *completely taken in by Congress . . . and by a sector of the Armed Forces.*

The swearing-in ceremony proceeded with the entire Joint Command present. *Life gave me the opportunity to be the first female education minister*, the new president asserted, and *the Ecuadorean people made me their first female vice president, and now, because of their Congress, I can hold the highest office in the land.*[30] There had been only three previous female heads of state in Latin America: Isabel Perón in Argentina, Lydia Gueiler Tejada in Bolivia, and Violeta Chamorro in Nicaragua,

Speaking from his stronghold in Guayaquil, Bucaram admitted that the struggle for power was temporarily over. He said he lost his authority to conspirators supported by the Armed Forces and claimed that *what is being formed in Congress is a civilian dictatorship. Remember me. In a short time these same people are going to beg me on their knees to come back.* Adum, his close friend and adviser as well as his energy minister, thought that Arteaga wanted the presidency from the day that she and Bucaram were inaugurated. *The vice president is a conspirator who seizes the president's balcony. I cannot tell if she works or not, but regarding conspiracy she labors efficiently and puts in overtime.* Learning of her swearing-in, Bucaram's reaction was simply, *She is an ambitious woman.*[31]

On February 10, Arteaga challenged the terms of her emergency appointment, saying she would not step down until the constitution was amended. Her choice for interior minister, Gil Barragan, argued that if Congress wanted to elect an interim leader it would first have to amend the constitution since it made no mention of a temporary executive. Arteaga was testing the agreement reached by her, General Moncayo, and Congress. Opposition to her announcement was fast and furious. The business community was among the first to respond with a strongly worded communiqué. The Chambers of Commerce and Industry, it stated, were shocked that Arteaga wanted to stay in office in violation of her prior commitment to observe the resolutions of Congress. The deputies were the most legitimate repository of the popular vote, and they had already spoken in favor of Alarcón. The businessmen declared that if she stayed on, she would do so without political or popular support and that consequently the nation would suffer enormous damage. More important than their stiff reaction was the fact that Arteaga's surprise revelation of her intention to retain power infuriated many people in Congress and within the Armed Forces. General Moncayo and others bluntly told her in private that for the well-being of the country she was to step aside. Months later, according to Arteaga, the general told her: *There was enormous pressure from Congress; all the leaders of Congress pressured him and the Armed Forces who thought that it was a political question that they* [Congress] *should decide, that the Armed Forces should not intervene.*

Painted into a corner, later in the day Arteaga reversed herself and pledged to resign, although she claims her sister Claudia received a call from millionare Luis Noboa, who offered to put whatever money was necessary at her disposal to fight Alarcón's election by Congress. She continued to focus on legalities. *I think they're preparing a new coup against the constitution,* she said on February 11, calling what was happening *a very dangerous precedent.* Before turning in her resignation, she issued a decree calling for a national referendum on whether the country's vice president should succeed the president if the top position were vacant. *I will return to the presidency of the republic only if that is the determination of the referendum,* she said, leaving the door open by plebiscite for her return to Carondelet. Meanwhile, she kept her job as Vice President. Arteaga resigned on February 11, after about seventy-two hours in office, and for the second time in six days legislators elected Alarcón as Ecuador's interim president. Congress refused to let Arteaga present her resignation in person: when she appeared at its doors, they were locked. *The cowards were afraid to listen to a woman,* she said.

THE DANCER

February 11, 1997. So-called for his political agility, Interim President Fabián Alarcón stands with military aide Colonel Lucio Gutíerrez, who looks in another direction. A little less than three years later, Gutíerrez helped CONAIE topple President Jamil Mahuad in January 2000. Alarcón was one of three presidents in three days.

The vote for Alarcón was 57 to 2, with five abstentions and one blank ballot, and he was sworn in as president, to serve until August 1998. He entered Carondelet at midnight. The U.S. ambassador, Canada's ambassador, and most European envoys did not attend Alarcón's swearing-in. Arteaga relates that Alexander phoned to tell her, *I am not going to go because this is a farce*. Reflecting upon her brief tenure, she concluded: *They usurped my power because I was a woman*.[32]

In 1998, Bucaram published his account of his brief administration.[33] The exiled former leader argued he was ousted from office in 1997 because he stood for change; he was an ally of the poor, an enemy of the oligarchy, and a fighter for privatization and modernization. He added that because he threatened to bring about currency stabilization, he was a threat to currency speculators as well as to those who gained from inefficient government enterprises. The oligarchy had its allies in the military, he explained, who were angered at his refusal to enter an arms race and at his insistence on peace with Peru. In particular, Bucaram blamed Febres Cordero, Nebot, and what he called their unprincipled political party of oligarchs.

The new president, Fabián Alarcón, was the son of Ruperto Alarcón Falconi, who in 1950 was elected to the Chamber of Deputies from Pichincha province. As his son was to do almost a half-century later, the elder Alarcón served as the head of that body. He ran as the Conservative Party (CP) candidate for president against Velasco Ibarra in 1952, and after his loss had a diplomatic career that included ambassadorships to Mexico and Spain until his death in 1968. Fabian was a lawyer by training and was named to the Quito city council during the period of military rule in the 1970s. In 1988 he ran for mayor of the capital and lost to Rodrigo Paz. Two years later, affiliated with the FRA, he won election to Congress and in 1991 was chosen its president with PRE backing. He again ran for mayor of Quito in 1992, was supported for the post by Quito's mayor, Sixto Durán Ballén, and suffered another defeat, this time to Jamil Mahuad. Two years later he was chosen deputy from Pinchincha province, once again supported by Durán Ballén (now president), and once more was tapped to head Congress. In 1996, during the second round of presidential elections, Alarcón endorsed Bucaram, aided his campaign in Quito, and put himself in a position to gain Bucaram's backing for reelection as president of Congress for the third time. During Bucaram's brief term, Alarcón made major efforts in the legislature for congressional and constitutional reform.

In February 1997, when Alarcón assumed the interim presidency, most observers thought Congress would essentially run the country until a new president was sworn in on August 10, 1998. From exile in Panama, Bucaram expected two conservative parties, the PSC and DP, to govern together. The deputies set a vague agenda in February that proposed to end most of the previous administration's austerity measures, take a new approach to the onerous foreign debt and the deficit, and enact both political and economic reforms to modernize the country. The new leader made it clear that the nation's condition was grim. Corruption, moreover, was a vital and highly charged issue: *The situation in which we have received the country is truly catastrophic. We have been at the verge of economic and social collapse. Because of that, one of the objectives of this government is to combat everything that signifies corruption.*[34]

Alarcón saw Arteaga's call for a national referendum on presidential succession as one that might strengthen his interim government, and while he embraced it, he chose to have it address more issues than his rival proposed. He scheduled a plebiscite for May 25 to provide institutional legitimacy and at the same time reform the constitution. Without the mandate of a popular vote and constrained by what he considered an inadequate economic plan devised by Congress, Alarcón considered the referendum his only chance to have more than a weak caretaker government. Although most people welcomed Bucaram's ouster, Alarcón recognized that both he and Congress were suspect for essentially usurping executive power. He and his small FRA, moreover, had a particularly weak public following. The party counted only two deputies in Congress. *The people,* Alarcón proposed, *without intermediaries, will tell the government what they want it to do in the coming months.*[35] They would speak by answering twelve questions. The first would judge Congress's removal of Bucaram, and the second Alarcón's designation as interim president. If voters said no on either one, Alarcón vowed to respect his rejection and step down. Other referendum queries included whether the vice president should remain in office, when to hold presidential elections, and whether to write a new constitution. The proposal set off another round of hostile and personal debate in a demonstration once again of the nature and extent of Ecuador's political fragmentation. Arteaga wanted the plebiscite limited to whether the vice president should succeed the president, and in late April she proposed that Alarcón call general elections for president, vice president, and Congress. He should renounce his office, she would relinquish her own, and the people could choose leaders directly as the constitution intended, rather than leave the choices up to Congress.

The reaction was mixed but largely negative. Nebot of the PSC rejected the notion as distracting and unproductive. Positive change was needed, he claimed, not political posturing. He broadened his condemnation of referenda to include Alarcón's, which he said mattered only to politicians and in no way would help solve the country's enormous problems. Deputy Franklin Verduga, also of the PSC, urged Arteaga to step down. She would never hold the office, he predicted, because she represented *el Bucaramismo sin Bucaram* (Bucaramism without Bucaram).[36] Verduga averred that the differences between Bucaram and Arteaga were never deep and that therefore she was partly responsible for the nation's sorry state. Deputy Wilson Merino, director of Alarcón's FRA, called Arteaga the Mother Teresa of politics and labeled her a tainted officeholder without authority to ask for anyone's resignation but her own. Alarcón, he explained, represented the people through their elected national Congress, while the entire Bucaram regime had just been junked by the near-unanimous will of the people.

Unwavering hostility to the plebiscite came from the PRE, whose director, Adolfo Bucaram, pledged a vigorous campaign to defeat it so that, he hoped, Abdalá Bucaram could return to Ecuador. The influence of the party was diminished by Bucaram's departure, and its weight was further reduced when thirteen legislators, mostly PRE members, were ousted from Congress in mid-April on corruption charges.

The referendum was conducted in May. Most propositions passed. Regarding the deputies tapping Alarcón to succeed Bucaram, 65 percent consented, 28 percent disapproved, and 7 percent of the ballots were blank. On the question of a convention to reform the constitution, the breakdown was 59 percent in favor, 27 percent against, and 14 percent blank. One of the queries addressed political fragmentation. Proposition 7 let voters decide whether to select legislators before or after they knew who would be the two finalists for the presidency after the first round of voting. The current system elected the deputies during the first round, at a time when no one knew who would square off in the runoff. Advocates for electing everyone at the same time believed the reform would help unify the government and reduce political divisions, and the proposition offered the choice of voting for congressmen in the first round of presidential elections, in the second, or in a special election if a second was unnecessary. The status quo won with 50 percent, while 30 percent favored either the second tier or a special vote. A constitutional convention was approved; on June 5, 1998, it issued Ecuador's nineteenth constitution. It raised the number of deputies in the unicameral legislature to 123, strengthened the executive by taking from Congress the power to remove cabinet ministers, and made it possible for state activities to be administered by public, private, or combined enterprises.

Elections were scheduled for May 31, 1998. Meanwhile, the economy continued its sluggish course. Alarcón inherited a budget in which 45 percent of the expenditures went to pay the foreign debt, while another 40 percent was allocated to operating expenses and already existing programs. Little discretionary money was left; when added to the domestic deficit, the figures made new social investment unlikely without raising additional revenue. Although the economy picked up in 1997, growing 3.5 percent more than it did during the previous year, the deficit and foreign debt both shot up. The first trimester of 1997, the one in which Bucaram was removed, gave the worst performance of the year. Sales fell 20 percent, production declined, and unemployment rose. Interest rates stood at about 48 percent, with inflation projected to run 37 percent for the year. Making matters worse, by April the price of crude oil was down to $13.54 per barrel, the lowest in a decade due to an excess of oil on the world market.

In April 1997, members of Ecuador's monetary council identified five ways to attack the deficit: raise existing taxes, create new ones, reduce spending, increase foreign debt, and privatize state-owned enterprises. Alarcón embraced all of the suggestions, with the goal of reducing the deficit by 6.7 percent in the next year. But the economy grew only .99 percent in 1998, and red ink continued to run. Yet like Bucaram before and Mahuad afterward, Alarcón maintained continuity in economic thinking. He insisted that the only way to put the economy on a sound footing was to correct recurrent fiscal deficits, impose new taxes, slash spending, modernize the state, put public enterprises in private hands, and encourage private saving and investment. Such was the theory. Because of political obstacles and public pressures, the practice was far from consistent.

Because of the lackluster economy, tax revenue fell 19 percent in the first trimester of 1997 compared to the same period the previous year. In mid-April, Alarcón sought to both spur consumption and reduce public frustration with price increases by slicing Bucaram's tax hikes an average of 30 percent. The alcohol increase was lowered from 63 to 20 percent, dark tobacco cigarettes from 48 to 15 percent, and light cigarettes from 103 to 67 percent. Electricity rates for 80 percent of the population doubled under Bucaram and were revised downward as well. The 10 percent tax increase on automobiles was left untouched.

The downward spiral of the economy worsened social conditions. Aimed at alleviation of the hardship endured by lower income groups, the Popular Fund established by the Ministry of Agriculture and Livestock subsidized basic food items including rice, sugar, powdered milk, and cooking oil. Deficit spending helped pay for underwriting this and other such programs, as expenditures in 1998 outpaced revenues by $470 million. Despite the Popular Fund, Alarcón, like his predecessor and successor, confronted a wave of protests and strikes.

Budgetary neglect fueled the discontent of workers in the public sector. The country's 15,000 public hospital employees, doctors, nurses, and administrative staff stopped work on April 14 to demand better salaries, collective bargaining, and a health budget equal to 12 percent of federal spending. The government countered with far fewer funds than those requested but pledged to devote 8 percent of appropriations to public health services. At the same time, schoolteachers made pleas for better salaries and voiced their opposition to the decentralization of education, while contractors pressed for payment of money owed for public works projects and businessmen insisted on rules from the government to give them greater flexibility with workers.

Local governments joined the chorus of demands, and like their counterparts in public health the municipalities demanded a greater allotment of federal funds. With few resources of their own, they felt stretched beyond their capacity to provide such essential services as garbage collection and street cleaning. Quito depended on the central government for 25 percent of its revenue, but some provincial cities relied on it for up to 90 percent. The Association of Municipalities explained that though local levies rose from 70 million sucres in 1992 to 380,000 million in 1996, they remained insufficient to cover basic needs, and most localities invariably registered deficits. The federal budget in 1997 tagged 4.5 percent of expenditures for the local units, but they continued to spend more than they took in.

Thirteen deputies, including Bucaram's brother Santiago, were expelled from Congress in early April, and by court order their property, bank accounts, and other assets were seized or frozen. The court took the drastic actions because of the disgraced men's participation in a computer scandal. The schemers set up a fictitious computer company, offered the sought-after machines to schools throughout the country, and then sold them at a 600 percent markup. Educational institutions made their purchases with public money coming from the extraordinary purpose fund of the national budget controlled by the president. Santiago Bucaram had fled to Panama in February, followed by Homero Fuertes, who skipped the country on April 17. Although the other eleven swindlers were not to be found, their photographs were posted at border crossings. Bucaram was also accused of spending $40 million in public funds to buy 500,000 student backpacks at a cost of $80 per bag for a subsidized school supplies program. The contract, given to a Colombian company without competitive bidding in late 1996, had called for 1.4 million schoolbags, but 900,000 were never delivered. No transaction, it seemed, was immune from corrupt politicians. And while fraud was hardly new, awareness of its details and depth were much greater than in past decades. Commentators claimed that public anger against political corruption was at an all-time high.

Sanctuary protected Bucaram in mid-April when the Supreme Court issued orders to arrest and imprison him on charges of stealing public money and engaging in illegal schemes for the enrichment of a cadre of his associates and himself. The exiled leader's response was that mental incapacity, not corruption, was Congress's reason for voting him out, and the arrest warrant with its accusations of stealing were designed to prevent him from participating in the forthcoming elections. Bucaram stood accused of taking 20 million sucres from the president's reserve expenditures account earmarked exclusively for internal or external security needs. He allegedly took the money on February 6, the day Congress voted him out of office, and gave it to his close friends and advisers, Abdum Ziade and Salem Kronfle. The pair, it was claimed, speedily converted the sucres to U.S. dollars. Bucaram insisted that the sucres went to purchase tear gas, but Police Commander Hinojosa countered that although he received such an allocation in early February, it was a fraction of the amount in question and left a substantial sum unaccounted for. The Supreme Court, an institution Bucaram and most analysts saw as highly politicized, alleged theft, ordered Bucaram's arrest, froze his personal bank accounts, and prohibited the sale or transfer of his property. The Supreme Court also issued an arrest order against Patricio López, former head of the state-owned oil company, Petroecuador, for illegal use of funds and accepting bribes.

Large-scale and widespread corruption in Guayaquil's customs house did not begin or end with Bucaram. Still, the near daily revelations about supposed swindles in the nation's commercial gateway shocked the public. Containers of merchandise frequently were allowed to pass through duty free, costing the country revenue while enriching corrupt officials. According to government investigators, those responsible included Bucaram and customs administrators Gustavo Flores Zapata, Carlos Hidalgo Villacís, Patricia García, and Víctor Suárez.

Peace with Peru was one of Alarcón's prime objectives. Ever since the 1995 border conflict, Ecuador was committed to friendship with its southern neighbor, as most observers saw either war or an arms race as far too risky and costly. Accordingly, representatives from the two countries met on April 16, 1997, in Brasilia in an attempt to resolve outstanding issues. The talks did not result in a final peace treaty but laid a solid foundation for Alarcón's successor to sign an accord on October 26, 1998. Meanwhile, a half-dozen candidates entered the race to succeed Alarcón.

6

Jamil Mahuad

There is excessive fragmentation of parties and the president of the republic, lacking a majority in the parliament, dedicates a great deal of his energies to trying to reach agreement with the opposition. The fragmented politicians condition their agreements on their immediate electoral interests. It creates a climate of constant instability.[1] Milton Luna, director of history at the Catholic University, likely reflected the view of most Ecuadoreans when he offered this analysis in February 2000.

The hope that the excessive political fragmentation would be reduced with the accelerated pace of modernization that accompanied the oil boom of the 1970s did not materialize. Old patterns to which Professor Luna referred managed to continue apace, as did the public perception that Congress adopted laws that attended only to the deputies' personal interests and to powerful groups, not to the national welfare. The last president elected before the oil bonanza was José María Velasco Ibarra, who took office in 1968 with only 32.84 percent of the popular vote, the highest tally in a five-candidate race. Without a runoff, Velasco Ibarra assumed office on the basis of garnering 280,350 votes of the 853,474 cast, about 14.2 percent of six million Ecuadoreans. In an effort to strengthen the state by requiring a stronger public mandate, the constitution was altered in 1978 to require a majority of the ballots cast in order to be elected president. For the rest of the century, this required two rounds of balloting in each presidential election.

In the two-tier elections from 1978 to 2000, the candidate with the highest percentage in the first round invariably went on to win in the second runoff election. In order to field a candidate, a party needed 5 percent of the vote in the last contest. That threshold, many thought, was far too low and constituted a major cause of political splintering in Congress. Candidates with relatively weak political support could make it through the first round to go on to win in the second. In 1988, to illustrate, Abdalá Bucaram received only 17.6 percent of the ballots in the first tier, but it was enough to put him in the runoff with Rodrigo Borja, who took 24.4 percent. The net effect was to force coalition government and to make it difficult for the executive and legislature to reach agreement on pressing issues.[2]

The composition of Congress, which was expanded to 123 members in 1998, was determined in the first round of voting. A host of political parties was represented, none of which mustered a majority of the seats. With the exception of 1988, when the ID had 42 percent, from 1984 to 1998 the PSC was the largest in the legislature, holding between 22 and 30 percent. Although Jamil Mahuad's DP outdistanced them in 1998, they remained strong after his election. (Congress in May 2000, moreover, would be as splintered as ever, with twelve parties dividing the 123 seats. Twenty deputies were elected at large and 103 were chosen by separate provinces for four-year terms. Nine political parties were represented, with Popular Democracy (DP) holding 29 percent of the seats; the Social Christian Party (PSC), 23 percent; and the Ecuadorean Roldista Party (PRE), 18 percent. The Democratic Left (ID) had 15 percent; Pachakutik-New Country (NP), 5 percent; the Radical Alfarista Front (FRA), 3 percent; the Conservative Party (CP), 2 percent; and the Popular Democratic Movement (MPD) and Concentration of Popular Forces (CFP) each had 1 percent. The center-right had the majority, consisting primarily of DP and the PSC as well as the FRA, the CP, and the CFP. The center-left was mainly made up of the PRE, ID, Pachakutik-NP, and the MPD.

As the May 31, 1998, election approached, six candidates sought the nation's top spot.

• Jamil Mahuad, Popular Democracy Party (DP), whose slogan was *Se que hacer y como hacerlo*: "I know what to do and how to do it." Mahuad emphasized his accomplishments as mayor of Quito. Like the other candidates, his plans for education, health care, housing, employment, and security were general. Unlike the others, however, he stated in clear terms that he would raise taxes and cut subsidies.

• Álvaro Noboa, Ecuadorean Roldista Party (PRE), whose slogan was *Adelante Ecuador, adelante!*: "Forward, Ecuador, Forward." Bucaram backed him from exile. Noboa's efforts were aimed at the lower classes, and most of his speeches emphasized housing and the need to rebuild large segments of the coast after El Niño devastated it. Although nominated by the PRE, he tried to present himself as an independent. He said he would declare a moratorium on payment of the international debt and stimulate economic growth by reducing taxes.

• Rosalía Arteaga, National Alliance Party (MIRA), whose slogan was *Empleo seguro, palabra de honor*: "Dependable Work, Word of

Honor." Arteaga distanced herself from the Bucaram administration and accented her devotion to the constitution, her experience with the nation's recent problems, and her ability to deal with adversity. Many observers noted that she did not address public policy issues or offer even a general approach to the economy but instead spoke of the recent "usurpation of the presidency."

• Rodrigo Borja, Democratic Left Party (ID), whose slogan was *Ofrezco experiencia y honestidad*: "I offer experience and honesty." Having served as president before (1988–1992), Borja spoke of his past administration's accomplishments. He was endorsed by popular general Paco Moncayo, who himself was elected to Congress as a national deputy from the ID. Borja opposed raising taxes until the economy was out of recession.

• María Eugenia Lima Garzón's Popular Democratic Movement (MPD), whose slogan was *Un pueblo que se levanta*: "A People Rising Up." She had served in Congress from Chimborazo province and said her candidacy was a chance for women to advance in politics since they made up over 50 percent of the population. Lima's emphasis was on women, class, and ethnicity, and she brought to the contest a Marxist conception of all three.

• Freddy Ehlers, Independent Citizens' Movement for a New Country, or New Country (NP), whose slogan was *Combatir la corrupción para un nuevo Ecuador*: "To combat corruption for a new Ecuador." With a promise not to rob, lie, or be lazy, Ehlers appealed to labor, rural workers, and the Indian population. Ehlers employed the huipala, the flag of the Indian movement. He stressed that corruption was the nation's top problem and traveled across the country, machete in hand, vowing to eradicate it. The Socialist Party backed the candidate, as did Pachakutik. Ehlers had never held office and fell short on specifics, but he said repeatedly that *he who wants to learn, learns*. Regarding his days as a student studying law at Quito's Central University, Ehlers admitted, *I left lacking fifteen days to finish, because I was ashamed to get the degree. I began to see what a pigsty the judicial sector was, as 90 percent of my classmates copied in class and everyone was given a degree.*

At the last minute, former Supreme Court head Carlos Solórzano opted out of the race. He did not endorse anyone and claimed the lack of money forced his decision. More important was the determination of Jaime Nebot of the PSC not to run. He had lost the last two presidential elections—to Durán Ballén in 1992 and to Bucaram in 1996—and few people thought he could win. He likely calculated that it was in his party's best interests to concentrate on controlling Congress. *We are going to govern from Congress*, he announced, as he launched his candidacy for deputy from Guayas. Nebot lamented that Ecuador seemed ungovernable, and he was frustrated that modernization failed to advance on any front. Nebot's late decision not to seek the presidency surprised and disappointed many members of his formidable party.[3]

None of the six candidates won the needed majority. The two who made it to the runoff six weeks later were Mahuad, who took 36.68 percent of the ballots, and Noboa, who gained 29.75 percent. Together they garnered 66.43 percent of the vote. Behind them were Borja with 14.66 percent, Ehlers with 12.97, Arteaga with 3.97, and Lima with 1.97. Mahuad's chances for victory in the runoff were boosted when on election night Nebot pledged to back him. Nebot's PSC and Mahuad's DP, two essentially conservative blocs, would face off against the center-left, led by Noboa and the PRE.

Forty-eight-year-old Jamil Mahuad was born in Loja but spent fifteen of his earlier years in Guayaquil before settling in Quito. The eldest of three sons of a Lebanese immigrant, his father died when he was seventeen. During Osvaldo Hurtado's administration (1981–1984), Mahuad served as minister of labor. In 1981 he joined the DP and was voted into Congress from Pichincha province five years later. Mahuad began serving his first term as mayor of Quito in 1992 and was reelected in 1996. As head of the nation's capital he was continually in the public eye and soon gained a reputation as a hardworking and competent administrator who followed through on initiatives. The mayor won particular praise for road improvements and for initiating a trolley car system in the increasingly congested capital. He had sought the presidency earlier, in 1988, but did not make it past the first round, finishing fifth in a field of ten and taking 11.5 percent of the vote. But the young candidate garnered more ballots in Quito and throughout the highlands than any other aspirant, and that strong regional approval was likely a major factor in his decision to run for mayor in 1992. After his failed presidential effort, Mahuad attended Harvard University and received a master's degree in public administration in 1989. In March 1997 Mahuad suffered a stroke, from which he recovered rapidly. Politically, most observers considered him conservative on the economy and center-left on social policy.

Mahuad's economic program began with a proposal to raise the value-added tax by 5 percent and to hike other duties as well. At its 10 percent level, Ecuador's value-added tax was the lowest in Latin America. The candidate pledged to earmark a significant percentage of this increase for reconstructing coastal roads damaged by El Niño.

Overall subsidies would be reduced because they went primarily to those who did not need them, but the poor would continue to be protected with underwriting for gas, electricity, and other essentials. The Central Bank claimed the cooking gas subsidy cost the state $170 million per year. The bulk of the benefit went to the relatively well off, while only 10 to 20 percent of it reached the two-thirds of the people classified as poor. Mahuad argued the same was true for telephone, electricity, transportation, and other subsidies. He also said he would create 900,000 new jobs and seek a new pact between labor and management that would make business more efficient through greater flexibility in hiring and firing. He criticized what he called a small circle of abusive labor unions.

The issue of whether to raise taxes and lower subsidies was the most significant substantive difference between Mahuad and Noboa. That fact plus political alignments during the campaign help illustrate the complexity of Ecuador's politics, replete with ideological and policy differences, personal rivalries, and jockeying for position. Mahuad advocated higher taxes, lower subsidies, and tougher austerity measures than his rival and was deemed the more conservative of the candidates. Borja, who came in third behind Mahuad and Noboa in the first round of elections, was considered to the left of Mahuad and closer to Noboa on tax issues. Borja insisted higher taxes made no sense during a recession, but he nonetheless endorsed Mahuad. Nebot and his PSC likewise opposed raising taxes for the same reasons as Borja but also backed Mahuad. The candidates and public alike were well aware of the gap separating election rhetoric and governing policies, as Bucaram's administration had demonstrated.

Mahuad believed that bringing expenditures into line with revenues was essential if Ecuador were to restructure existing international loans and win new ones. The country could not continue to pay its foreign debt obligation, he believed, and would have to seek new terms with lower rates and longer payback periods. Privatization would be part of a rigorous economic plan; it would stimulate the economy, raise more investment money, increase efficiency, create new employment, and reduce government spending. As part of his social program, he urged building 50,000 houses per year and asked municipalities to donate both the land and basic services. Although his health proposals were vague, Mahuad advocated free daily vitamins and adequate food for pregnant women living in poverty, supported the principles of universal health care coverage, and favored decentralizing the delivery system. He planned to seek international financing for a school breakfast program and to decentralize schooling.

Education was essential, Mahuad argued, as it was the only chance aside from winning the lottery that those in dire straits had of escaping from poverty. The poor, he pointed out, were in the majority, and they needed education the most. The candidate claimed corruption in the customs houses could be curtailed by adopting the Peruvian system that divided imports into three categories: those believed to have no, little, or substantial problems in paying customs duties. He urged verification of cargoes before they were unloaded to help eliminate irregularities in paying the proper duties.

Mahuad's opponent was 47-year-old Álvaro Noboa of Bucaram's PRE, Ecuador's richest man, who was making his first bid for public office. Álvaro was the son of Luis Noboa Naranjo, architect of the largest, richest, and most powerful enterprise in Ecuador, Noboa Banana Exports, which owned 90 percent of arable land in Los Rios province south of Guayaquil. The fifth largest banana company in the world, competing with Del Monte and Standard Fruit, it controlled both production and distribution and owned its own fleet for shipping to overseas markets. Diversification into other export crops and banking, moreover, ensured that the company withstood short-term crises in the industry. When Luis died in 1994, his empire was divided, with his widow receiving 48 percent of the fortune and his five children sharing the remainder. In 1998 presidential aspirant Álvaro Noboa's net worth was said to be $900 million.

Noboa claimed to have operated over 100 family enterprises. Following his family's custom, he contributed substantially to the presidential campaign of 1996, donating to Bucaram, and after the triumph of his candidate of choice he was named to the country's Monetary Fund. His politics were a switch from his father's, as the elder Noboa had backed Bucaram's archrival and enemy, Febres Cordero of the PSC. Álvaro backed Bucaram's dollar-conversion proposal and remained a loyal supporter to the end. A lawyer and a millionaire dozens of times over, he claimed his policy proposals made him the *leader of the poor*. Many observed that Bucaram had done much the same and that both men were aided by a large cadre of multimillionaires. When Noboa entered the first round of elections, polls gave him less than 10 percent support, but the numbers continued to grow. He came within one percentage point of winning the presidency.

The candidate's favorite phrase was *¡Adelante Ecuador, adelante!* To take the country forward, in contrast to Mahuad he opposed raising the value-added tax. Moreover, he insisted that overall taxes had to be lowered to give the economy a desperately needed boost. Subsidies would be maintained for the poor, and if necessary the least fortunate would be given cards to purchase cooking gas and basic services at reduced rates. His plan, he assured voters, would encourage domestic spending and investment, with the result that upward economic movement would attract ever-increasing foreign capital. Tax reduction would be the initial key to stimulate the economy, which in turn would generate more revenue.

Even so, Noboa doubted Ecuador could continue to pay on its foreign debt, and he advocated a moratorium on loan repayment. At the same time, and without commenting on the possible linkage between a moratorium and additional borrowing, Noboa promised to win international loans to lay the foundation for a new economy. The subsequent economic dynamism, he was confident, would soon dwarf the country's current $14 billion foreign loan obligation. Noboa was sure that new loans would be forthcoming if Ecuador was perceived as truly serious about creating a dynamic economy. He pointed to Taiwan and argued that since it was a smaller country than Ecuador, had about the same number of people, and possessed far fewer natural resources, the fact that the island had $100 billion in monetary reserves proved that Ecuador could do the same.

Noboa pledged to raise the minimum wage, noting that no one could live on their earnings at the present level of about $53 per month. He promised 200,000 houses to replace those destroyed by El Niño at an average cost of $3,000 per unit. During the campaign he passed out T-shirts and toothbrushes by the thousands. He emphasized preventative medicine as the cornerstone of the health-care system and said he would seek integration of public and private health providers. Further, Noboa believed that national technical schools should be geared to the needs of the economy and called for financial aid for needy students to attend colleges and universities. On the local level, to ensure accountability and quality instruction, he wanted primary and secondary schools to create directorships composed of teachers and parents. To tackle corruption in customs, Noboa was willing to give the Armed Forces a significant measure of control and was confident that his proposal to lower taxes would aid in reducing improprieties. Finally, to distance himself from Bucaram, he promised that no member of his family would be part of his government.

Noboa emphasized the obvious: He was far too rich to steal. During the campaign the candidate may well have set a world record for the amount of personal money spent to run for office. Noboa's outlay was at least $8 million in the first round, and a minimum of $2 million in the second. He spent at least $10 million in a country of 12.4 million people while another wealthy entrepreneur in the United States, Steve

Forbes, spent $40 million on both of his campaigns for the presidency in a nation of 240 million. Mahuad was outspent two to one in newspaper advertisements, and Noboa also purchased more radio and television advertising. Some people speculated that because of his occasional rapid and uncontrolled leg movement, a nervous system disorder, he avoided invitations to television interviews and debates. [4]

Noboa claimed a Mahuad government would be no different from that of Alarcón, and he offered himself as a clear contrast to both men. As the election tightened, negative campaigning increased. With five days left, for example, the press reported that Mahuad, who was known to have a daughter from a prior marriage, was the father of a nine-year-old boy, Jamil Pedro Mahuad Romero; that the child's mother, María de Lourdes Romero Castañeda, had sued him for child support; and that only after blood tests proved paternity did he comply with his legal obligations. The revelations apparently had little if any impact on the election. The Noboa campaign made allegations that Mahuad's brother Eduardo was tied to the narcotics trade in 1993, but that accusation also failed to do damage. Mahuad fought back, saying his opponent was an unrealistic demagogue, as demonstrated by his promise to provide 900,000 new jobs. He claimed he himself had made few promises, *but they are ones I can keep. Those that see themselves losing offer everything to win votes, but they know that there is no way they can comply with what they offer*. Mahuad insisted it was indecent to raise the hopes of the poor by promising what they would never receive.

The major difference between the candidates was that Mahuad offered belt-tightening and higher taxes while Noboa pledged more government spending and lower taxes. For the most part the business community backed Mahuad, as did Jaime Nebot's conservative PSC. Nebot throughout the campaign tied Noboa to the exiled former president. *It is Bucaram, it is the Ecuadorean Roldista Party*, he maintained. *I cannot vote for robbery, for repeated theft, for incompetence, for those who would destroy the country, and because of that I will vote for Mahuad*. Yet Rosendo Rojas, a deputy and member of Pachakutik, spoke for many people when he said that Mahuad's triumph would mean a consolidation of neoliberalism. The deputy dubbed Mahuad *conservative in part, liberal in part, but profoundly neoliberal*, and cast a blank ballot in the presidential contest. [5] For its part, CONAIE remained neutral.

At the outset, Mahuad had a triumphant air, which most observers said turned to paralysis within weeks and to near panic toward the end. Commentators claimed he failed to raise popular enthusiasm because his speeches were cold, overly analytical, and without emotion. Moreover, in the runoff Mahuad did not move significantly closer to either the PSC on the right or to the ID to the left of center. As a consequence, he failed to arouse either of the major sectors that had hoped to pull him closer to its point of view. Still, most predicted that, if elected, Mahuad would ally with the PSC, which was ideologically close to his own party, and that a progressive force would likely form on the left, consisting of the ID, NP, Pachakutik, and the socialists. The analysts proved largely correct.

On election day both candidates claimed victory on the basis of exit polls, but the votes went to Mahuad by the thin margin of 51 percent: 2,235,685 to 2,124,747. Foreign observers who monitored the election thought the vote was fair; nevertheless, Noboa quickly alleged fraud. In an open letter to the nation's Armed Forces, he wrote: *This serious crime against democracy—this electoral fraud—must not be consummated.*[6] The defense minister indignantly replied that soldiers had supervised the elections and that all the candidates except the loser recognized that they had been honest.

On August 10, 1998, Jamil Mahuad was inaugurated president for a term to expire in January 2003. The image that he projected stood in sharp contrast to that of the last popularly elected leader, namely, of a capable, honest, and youthful public servant. Yet like his predecessor he failed to pass the tough measures needed to stabilize and modernize the economy. His attempts at belt-tightening met with repeated protests in the streets, which affected both his course and that of Congress. Reflecting on the executive's relationship with the legislature, *El Comercio* editorialized that with apparent support for approval, he would send an initiative to Congress only to watch a seemingly never-ending process of modifications by various political parties and coalitions. Vladimiro Álvarez, who began as Mahuad's minister of education and ended as his minister of the interior, summarized the president's economic program: *cutting public spending, reducing the bureaucracy, increasing taxes, enforcing the collection of existing taxes, improving the process of tax collection, broadening the base of tax contributors, eliminating subsidies, elevating the price of fuel and the charges for electricity consumption and telephone use . . . or creating new taxes.*[7]

Several of Mahuad's problems were not of his own making. The 1998 budget assumed oil prices, which sold at a discount over West Texas Intermediate oil, of $16 per barrel; they fell to $9 per barrel. Alarcón had spent the entire 1998 budget before he left office in August, leaving Mahuad with nothing for the remainder of the year. Unlike his predecessors, the new president did not have so-called discretionary funds with which to grease palms and sway deputies to vote for his programs; the 1998 constitution had banned them after their misuse by former Vice President Alberto Dahik and Bucaram. The harsh winds and rains generated by El Niño had devastated the coast's infrastructure and crops. When as a consequence bank loans could not be repaid, the nation's poorly regulated financial system was shaken and eventually fell apart.

Basic patterns of presidential action and public reaction were set early on. As urged by the IMF, Mahaud planned to reduce expenditures by cutting state subsidies across the board. After five weeks in office, on September 14 he presented his first economic proposals, which raised fuel and electricity prices 410 percent. The heavily subsidized cost of public transportation on which the vast majority of Ecuadoreans relied increased 40 percent. The cost of living rose for the vast majority of people because wages failed to even come close to keeping up with the nation's soaring inflation. To help mothers living in a state of dire poverty, Mahuad urged that each one be given a small sum. The predictable public reaction to Mahuad's measures came on October 1, when the first nationwide strike was called against his economic austerity and free-market policies. The demonstrations continued and grew stronger through the end of the year.

The first quarter of 1999 saw major changes in monetary policy and sporadic protests against price hikes. On February 12 the government allowed the sucre to float freely and find its own market value in relation to the dollar, as opposed to having its worth set by state financial institutions. The sucre fell sharply when the Central Bank eliminated the controls which set the exchange rates, that had been in place since late 1994. The loss of foreign currency reserves, Mahuad explained, had forced the action. In the face of swelling hostility to his monetary moves, reductions in subsidies, and proposals for additional tax hikes, on March 9 the president was forced to declare a state of national emergency. Public criticism mounted two days later when he raised the value-added tax from 10 to 12 percent and increased gasoline prices 174 percent.

As the president attempted an austerity program, five issues besides problems with Congress, challenged the government. These were: (1) the peace with Peru; (2) the banking crisis and corruption; (3) Plan Colombia; (4) demands for Mahuad's resignation; and (5) the plan to convert the sucre to the dollar.

Since Ecuador and Peru declared their independence from Spain, each of the neighboring nations claimed the vast and largely uninhabited jungle area that separated them. Ecuador's attempts to secure the disputed land were confined largely to diplomatic efforts, relying on colonial precedents and principles of international law. And while each claimed the weight of history, Peru had substantial advantages, as Ecuador learned when its neighbor resorted to force to settle the controversy.

Over one-half century before Mahuad took office, in 1940 Peruvian troops mobilized along the border. The United States, Argentina, and Brazil offered to mediate. Weaker Ecuador accepted but stronger Peru refused, and following a series of skirmishes between border guards the more powerful of the contenders, without a formal declaration, launched a 21-day war from July 5–26, 1941. With a population twice the size of Ecuador's and a military budget twelve times as large, Peru pitted a force of about 13,000 invaders with tanks, artillery, and air power against fewer than 1,800 Ecuadoreans who defended the frontier. Ecuador's southern provinces of El Oro and Loja were easily taken. After occupying more than the disputed territory, Peru then bombarded a few towns in the far south and imposed a blockade on Guayaquil. Although small and poorly equipped, the Ecuadorean Army resisted near Zarumilla in El Oro province. Ecuador's border forces were outnumbered thirty to one, and most of its machine guns failed to fire for lack of oil. Only a few volunteer units arrived as reinforcements and the Ecuadoreans were speedily routed. After that, the Peruvian invasion went largely uncontested. President Carlos Arroyo del Río (1940–1944) feared a coup d'état and as a consequence kept his best troops in Quito. At the urging of other nations of the hemisphere, Ecuador's leader placed near-exclusive reliance on diplomatic discussions. In control of El Oro province and poised against Guayaquil, Peru issued an ultimatum: its forces would evacuate only if Ecuador agreed to a final and binding accord. If Quito did not propose an acceptable treaty within six months, then Lima would impose a solution by force

Representatives from Chile, Argentina, Brazil, and the United States who were planning to meet in Brazil in early 1942, to discuss entry into World War II, added the Andean conflict to their agenda. The final result was the Protocol of Río de Janeiro of January 29, by which Ecuador lost more than half of what it considered its national territory and virtually all of the easily navigable portion of the Oriente's rivers. Had Quito not accepted, Lima doubtless would have proceeded to occupy even more land and inflict heavy damage on Ecuador's commercial hub of Guayaquil. Peru's invasion came about five months before Pearl Harbor, and the Protocol of Río was signed about two months after the Japanese attack. The United States and other countries, given their focus on war against Germany in Europe and Japan in the Far East, likely would have done little to stop Peru.

After the disastrous defeat, humiliated Ecuadoreans acknowledged the loss of about half of their territory. Resentment was high. In 1945 a constituent assembly voted to confiscate exiled former president Arroyo del Río's land and possessions and strip him of his rights as a citizen, and asked that he be tried for treason and put to death. In 1960, President Velasco Ibarra decried the Río Protocol and declared it null and void. In the face of his and others' fulminations against the treaty, from war's end until Mahuad signed a peace pact with Peru in 1997, tension prevailed along the 1,000-mile border as unsuccessful efforts were made to draw the final boundary. Violence broke out in 1978, 1981, 1983, and again in 1984, leaving several hundred dead. With Durán Ballén heading Ecuador and Alberto Fujimori in charge of Peru, on January 26, 1995, warfare suddenly flared along a 48-mile area of contested border. There, Ecuador had established a number of permanent and for a time undetected bases, trenches, and land mines in the Cenepa River region east of the Cordillera del Condor in eastern Zamora Chichipe province. Battles raged for thirty-four days, and the fort of Tiwintza became the symbol of Ecuador's national sovereignty. Groups of Shuar and Achuar Indians in the remote and swampy region were the people most directly affected. At this time nationalist fervor rose in both countries, replete with marches, demonstrations, and protests. When Ecuador shot down Peruvian aircraft and Peruvian infantry streamed to the border, thousands of troops were mobilized on both sides.

Peru got somewhat the worst of it during the brief flareup. Ecuador held its own despite inferior numbers, arms, and equipment. The commander of the Ecuadorean forces, General Paco Moncayo, became immensely popular and after the fight over the strip of border was dubbed the "Hero of Cenepa." On February 17, 1995, with hundreds dead on both sides, a cease-fire was signed in Brazil providing for demilitarization and international observers along the disputed border. There were reports that Peru lost 1,000 soldiers and a considerable number of aircraft, and that Ecuador had only thirty-one casualties. In contrast to the humiliating defeat of 1941, Ecuadoreans now felt they had regained their national dignity in standing up to Peru.

Neither Peru nor Ecuador could afford an arms race or a major war; both faced serious internal problems and wanted to eliminate the risk of future hostilities. In an allout war, Peru's President Fujimori promised there would be *saturation bombing* by U.S.-made A-37 jets, French Mirage fighter planes, and Russian-made Sukhoi 22 bombers, and that *1,000 counterinsurgency troops would be engaged, all with jungle combat experience*. Peru, analysts and casual observers believed, would win.[8] Moreover, Ecuador had spent more than $900 million on the 1995 engagement, a formidable sum. During the conflict, workers and employers alike had contributed two days' pay to help finance it. When a peace accord was reached in late 1998, Ecuador still owed foreign lenders around $400 million. Meanwhile, Peru achieved both economic recovery and significant success in containing the guerrilla movements that for over a decade had disrupted its economy and social structure. During the late 1990s the efforts of the Shining Path and the Revolutionary Túpac Amaru Movement appeared greatly diminished if not contained. Conflict with an ever more stable and stronger Peru would be both costly and risky. Under these conditions, and confronted with an economic crisis, Jamil Mahuad was determined to put an end to the threat of war.

The main issue in bilateral negotiations was the fate of Tiwintza, the small outpost on the border. It went to Peru. The treaty established national parks on both sides of the border, demilitarized the boundary, and included trade agreements. Ecuador obtained free rein over a section of Peruvian territory as part of its navigation rights to the Amazon, including two free-trade zones of 150 hectares each along the river. The Inter-American Development Bank quickly offered support for the peace plan and pledged at least $500 million in loans for border development. In Brasília on October 26, 1998, the peace papers were signed that terminated fifty-six years of tension and hostility. It was the high point of President Mahuad's tenure.

Not all Ecuadoreans embraced the accord with Peru. A number of officers believed their president's concessions made the sacrifices at Cenepa meaningless. General Mendoza later wrote, *A good part of the nation was not in agreement* with the peace treaty; indeed, Mahuad had *committed grave errors in negotiating it*.[9] On the other side was former president Bucaram, who called the conflict with Peru *a stupid war that has no reason for being. Ecuador can never advance, nor can Peru, while their Armed Forces continue on a war footing. One cannot seek peace by buying arms and trying every day to become stronger to defend the frontier*.[10] As a dividend of peace, Mahuad reduced military spending, including a 40 percent reduction in the number of those obligated to military service. The Armed Forces now were to focus on fighting crime, narcotics traffickers, and guerrillas, and about 25 percent of their personnel were assigned to internal security.

Colonel Fausto Cobo, head of the nation's War Academy, summed up the military's reaction to the peace accord and its aftermath:

PEACE WITH PERU

October 26, 1998. Presidents Jamil Mahuad of Ecuador and Alberto Fujimori of Peru reach a peace accord in Brasília that ended decades of conflict and uncertainty between their neighboring countries. Mahuad was the son of a Lebanese immigrant to Ecuador; Fujimori's father was a Japanese immigrant to Peru.

> *After the peace accord was signed, the Army was left without leadership. The paradigm of war and the arm was left hanging in the air. Nobody had the capacity to reorient. Everyone spoke of an Army that had won the war, but the soldiers felt frustrated, the officers unmotivated; the president had taken measures against the Armed Forces, had reduced the number of conscripts and the budget, and the strategic and operating capacity was at truly low levels. The high command was happy in their rosy cloud, while the officers and troops had confidence only in their colonels, and we were the only ones who maintained, or tried to maintain, motivation in the Armed Forces.*[11]

As the 1990s started, democracy and free-market reforms swept Latin America, but scandalous activities in banking and finance at the end of the century undercut public confidence. In El Salvador the privatization of banks ended up with President Alfredo Cristiani winning ownership of the largest one for himself. In Guatemala the state selloff of the telephone company was being redone because the sale was tainted by corruption. In Mexico former president Carlos Salinas de Gortari, who started the 1990s as a reformer, finished the decade in exile and accused of stealing millions. In Argentina, Carlos Menem's decade of free-market reforms ended with most of his ministers having resigned after corruption scandals. Corruption had become a way of life, with politics as a vehicle for personal enrichment. Yet at the outset of his administration, Mahuad enjoyed a reputation for professionalism and honesty. Early in his tenure that opinion unraveled in the face of the country's financial crisis. The financial system collapsed in 1998–99, and its cost to the country equaled about 22 percent of the gross domestic product, making it one of the worst banking crises of modern times. The names Peñafiel, Isais, and Aspiazu soon filled the headlines of the nation's newspapers along with the institutions they headed: the Banco de Préstamos, Filanbanco, and the Banco de Progreso.

Mahuad considered it essential to take controversial steps to save the nation's shaky banking system, which for over a decade had posed a serious problem for the economy. There were numerous causes for the collapse, including the weakness of the Superintendency of Banks. The 1994 Law of Financial Institutions, according to one of Mahuad's advisers, followed the principle that well-administered banks would survive and that those that were not would have to close their doors. Ecuador had never developed truly independent regulatory agencies, as the regulators were drawn from the ranks of the regulated. Just as political parties controlled the Supreme Electoral Tribunal that was entrusted with certifying new political parties and counting votes in elections, so too did bank superintendents come from the banking community and customs administrators from the world of exporters and importers.

The justification for the arrangement was that only those with first-hand experience could offer competent guidance. Bankers ran the nation's monetary board and knew well in advance of any devaluation of the currency. As a consequence, they made fortunes simply by knowing when to buy or sell sucres. When their institutions faltered, as they did in 1984 during Osvaldo Hurtado's administration, bankers relied on the government to bail them out. If loans made by U.S.

financial institutions to Ecuadorean banks could not be repaid, the government assumed the obligations in dollars with the condition that Ecuadorean bankers reimburse the state in sucres. Currency devaluation followed; taxpayers picked up the stable bill owed in dollars while bankers saw their debts reduced by devalued sucres. In December 1998, *Vistazo* magazine reported the following conversation with an unnamed official from Filanbanco. When asked why the bank had opened so many branches throughout the country, he replied: *The longer the tail of the dragon the better, because he is more powerful. That gives good results in the banking business, because a government does not exist that can resist helping in moments of crisis.*

Many banks had gone broke after a period of rapid expansion and competition. Filanbanco claimed that 44 percent of its portfolio consisted of secure Class A holdings, but a subsequent audit showed only 8 percent. At the other extreme, while the bank maintained that its weaker Class D and E holdings made up a mere 13 percent of its portfolio, they were actually around 37 percent of the total. Other major problems were the high concentration of inadequately secured loans, many of which went to businesses in which the bankers or their relatives had ownership interests. From October 11, 1994, until August 24, 1998, four banks and seven other financial institutions collapsed. Ecuador's economic crisis of 1998, according to Jeffrey Franks of the IMF, would have caused problems for banks in any country, but given the wobbly loans and poor supervision of the banking system after 1994, it caused a crisis.

On August 25, 1998, the Banco de Préstamos was pronounced insolvent. Three-fourths of the institution was owned by the Peñafiel-Salgado family, prominent players in the oil industry, whose head was Alejandro Peñafiel. Upon learning of the bank's troubles, some of its major clients, including the National Police and the Social Security Institute, made massive withdrawals. Filanbanco, a bank directed by the brothers William and Roberto Isais, joined the Banco de Préstamos in asking for government help. President Mahuad as well as a growing number of Ecuadoreans were shaken. More than his predecessors of the last decade, all of whom suspected serious flaws within the banks, Mahuad probed deeply and was shocked by the breadth of the financial crisis. He was assisted by multilateral financial organizations, including the IMF, World Bank, the Inter-American Development Bank, and the Andean Development Corporation. The World Bank proposed the creation of the Deposit Guarantee Agency (AGD), established in December 1998 to give depositors confidence and avoid more bank closings. The move was deemed counterproductive by those who believed that monetary advances to cover deficits made bankers worry less about risk in making major loans for the purpose of recovering recent losses more rapidly. One of Mahuad's advisors put it bluntly: *Why make a major effort in managing a bank if the state will ultimately pay depositors?* Others likened the AGD to an insurance company set up to insure terminally ill patients.

INDIAN PROTESTS

March 1998. Indigenous people from the Nizag community and others march against the high cost of living in Riobamba, the capital of Chimborazo province. Indígenas constitute the majority of Chimborazo's population, but their numbers nationwide are a matter of controversy. The Indian movement CONAIE claims 40 percent, while others set the figure as low as 16 percent.

The AGD was financed in part by the tax on the circulation of capital, a levy of 1 percent on all deposits and transfers. The tax, adopted by Congress and signed by Mahuad in late 1998, had been introduced by the PSC. The rate was pronounced too high by most international lending institutions, and, as predicted, it constricted financial transactions and dampened growth. Many depositors reacted by keeping their money in safe deposit boxes or at home rather than pay the new tax. Some said it created a new institution, The Mattress Bank. With fewer depositors, banks had less money to lend. The economic downswing and banking crisis had another impact: they greatly accelerated the acquisition of stable currencies, above all U.S. dollars, which eroded the sucre and fueled inflation.[12]

Although in November 1998 Mahuad was quoted in *El Telégrafo* newspaper as saying *there is no problem in the country's banking system. I tell you emphatically, there is no problem*, by late 1998 he came to the conclusion that he had to stem the slide.[13] In March 1999 he ordered a week-long bank holiday, then partially froze deposits to limit the amount the nation's more than 4 million savers could withdraw from their accounts. Initially he hoped to keep the banks closed for one day while he announced a strong monetary program, including convertibility of the sucre. Instead, Mahuad and his major economic advisers, including Minister of Finance and Public Credit Ana Lucía Armijos and General Secretary to the President Ramón Yulee opted to steer a middle course between doing nothing and imposing shock treatment. It became apparent that once the banks were closed, any reopening would result in massive withdrawals from fright-filled depositors; and if too much money were taken out of the banks, more institutions would collapse. Between $3.2 and $3.8 million were frozen on March 11. The president did not propose convertibility. Mahuad's Minister of the Interior Álvarez justified the measure by declaring: *The government froze the deposits of thousands of Ecuadoreans in order that in reality they would not lose all of it forever.*[14]

The limitation on withdrawals instantly made Mahuad extremely unpopular. All of the nation's newspapers condemned the action, which most called an unconstitutional confiscation of private property. Former president Hurtado, the founder of the president's own political party, DP, began to criticize the government. Vice President Noboa, who was frozen out of decision making from the outset by Mahuad, later said: *After the banking crisis, the people saw that there was money to rescue rich bankers, but not to invest in necessities for the poor.* Adding to the outrage of depositors was the fact that the banks benefited from use of the frozen accounts and paid less interest on them in comparison to the rising interest rates they charged borrowers.

The head of the Armed Forces was appalled. General Mendoza later wrote of widespread public anger aimed at Mahuad's attempted bank bailouts:

> *Giving money to Filanbanco was a flagrant injustice in the face of the country's poverty and the elimination of gas and electricity subsidies. Citizens asked how the government could give public money to a private bank managed by a family? While bankers were given millions of dollars, President Mahuad exhorted Ecuadoreans to limit themselves economically, just as he asked the Armed Forces to participate in budgetary sacrifice . . . for*

> *him the reactivation of the economy would begin with the rehabilitation of the broken banks. One could detect that the government's economic plan gave priority to unconditional support for banks, with the absolute abandonment of meeting other public necessities.*

The administration, Mendoza privately concluded, *was the prisoner of an economic oligarchy and of corrupt bankers.* [15] Álvarez later justified the freezing of the accounts by saying it was necessary *to impede the absolute closing and collapse* of the banking system, which was *produced by many years of not having appropriate legal mechanisms for control and intervention.* [16]

The attempted rescue of the banks severely damaged President Mahuad's credibility, not only because of its considerable inflationary cost but also because many of the directors fled the country with large sums of public money advanced to restore the nation's financial health. Some took funds from their own failing banks, and some from credits extended by the Central Bank to help private institutions that had liquidity problems. Liquidity, however, was not a transitory problem, and the money advanced was insufficient, even though help from the Central Bank, it was hoped, would slow or stop the withdrawals by nervous investors and allow accounts to be unfrozen. Mahuad's political difficulty was that while millions of dollars of state funds were advanced to private banks under the pretext of saving the system, he had close ties to many of the prominent bankers who had left the country.

One director who did not make it abroad was the former president of the Banco del Progreso, Fernando Aspiazu, who was jailed on July 20 and accused of embezzling more than $180 million belonging to clients of the Banco de Préstamos. The charges included his failure to transfer to the government $7 million derived from the 1 percent tax. His Banco de Progreso was a major financial player and accounted for about 31 percent of all bad bank debt, which along with two other banks accounted for around 60 percent of the total. Aspiazu put a large portion of his personal assets in a trust fund created just as the Banco del Progreso started to sink. Government officials asked the courts to abolish the trust and argued they could secure from it at least $50 million to compensate Aspiazu's clients.

Matters were made worse when chief government attorney Mariana Yépez was accused by opposition politicians and much of the press of dereliction of duty and of attempting to protect allegedly corrupt bankers, notably the Isaias brothers. Filanbanco, their failing enterprise, was the first of many that asked for help from the Central

Bank. Critics claimed Yépez delayed prosecution for months after receiving the results of a special commission's investigation of Filanbanco. She then ordered the police not to detain the bank's executives when they were about to be indicted on several charges of corruption. The decision reputedly gave the Isaias brothers time to flee the country. In June 2000, Banking Superintendent Juan Falconi would issue his final report on Filanbanco, which found it improperly used money lent by the government to help it avoid bankruptcy. Contrary to loan conditions, just prior to being taken over by the AGD, almost $100 million was transferred from Filanbanco headquarters in Guayaquil to its offshore operations, many of which reported nonexisting enterprises among their holdings and had made loans with insufficient collateral.

Another banker who made it abroad was Alejandro Peñafiel, the former owner of the Banco de Préstamos, one of the first fifteen of thirty-five banks to go bankrupt. In August 1998 his institution was dissolved due to a lack of liquidity and failure to comply with requirements set by regulatory authorities. Peñafiel promptly fled Ecuador while the accounts of many depositors were frozen. Ecuadorean judges believed him guilty of corruption and fraud and wanted him prosecuted. The banking problem involved many others, as later documented by Falconi's November 2000 presentation to Congress in which he condemned what critics called the incestuous relationship between business and politics. Falconi offered evidence linking fugitive bankers Peñafiel, the Isaias brothers, Nicolás Landes, and others to an array of congressmen and political parties.

Mahuad froze from $3.2 to $3.8 billion, a huge amount in a nation where total annual output was $14.5 billion. The impact on savers was enormous; inflation kept advancing, wages did not keep up, and many people were now left without recourse to their savings that lost value almost daily in the inflationary economy. In April 2000, when the accounts were finally released, Luis Orden, a pharmacy owner in Baños, Tungurahua province, took out his checkbook and showed it to a foreign visitor. Like many others in Ecuador, his life savings, which had been put beyond his control by the government for a little over one year, had fallen from $3,000 to under $270 as he watched in horror.[17]

Opposition to Mahuad's directives was reflected in his approval rating, which fell from 60 percent in August 1998 to 22 percent six months later in February 1999. As the recession worsened, inflation increased, subsidies were cut, and millions of dollars were given to banks in an effort to shore up the nation's financial system. The Indian movement, a broad spectrum of labor unions, a variety of non-governmental organizations (NGOs), and others launched a nation-wide general strike and mounted demonstrations across the country. They opposed the 2 percent hike in the value-added tax to 12 percent, the increase in fuel and electricity prices, and the partial freezing of bank accounts. From March 12–27, taxi and bus drivers, with the help of students, workers, and thousands of Indians, erected roadblocks in the major cities. Their actions disrupted trade, commerce, and traffic and sent a clear signal of the scope and intensity of hostility to the belt-tightening measures. The president reacted by imposing a state of emergency, one of many declared throughout 1999 as protests mounted. With the rationale of combating crime, martial law remained in effect in Guayaquil from March through July. Rumors spread that Mahuad had considered and rejected a suggestion by Minister of Foreign Relations Benjamín Ortíz that he imitate Peru's President Fujimori's actions of 1992 and dissolve Congress, assume extraordinary powers, and immediately implement his economic program. PSC deputy Jaime Nebot affirmed that Ortíz had made the recommendation. Armed Forces head General Mendoza later wrote that throughout Mahuad's administration, he and other military men had *continually given speeches about the need to maintain the legal order and the democratic system.*

CONAIE's leader, Antonio Vargas, called for Mahuad to either modify his austerity measures or resign. He urged the movement's members to close roads, call newspapers, radio, and television stations, and wear black arm bands to express their indignation. If necessary, Vargas warned, there would be an uprising. Indians would march from the provinces to the capital where they would stage demonstrations until their demands were met. CONAIE insisted on the revocation of Mahuad's austerity measures, the immediate release of bank funds, and the arrest and prosecution of bankers guilty of mismanagement. Vargas urged that in the future major legislative proposals be put before the people in plebiscites instead of being imposed by unrepresentative regimes. Further, CONAIE advocated a more progressive tax system, elimination of loopholes, and severe punishment for tax evaders. It insisted that Mahuad renegotiate the foreign debt and attempt its modification if not forgiveness. Instead of paying high interest, CONAIE claimed the state should create a development fund for indigenous people. And rather than privatize vital sectors of society such as education, health care, electricity, telecommunications, and oil, it argued that the state should maintain a central role to seek social equity.

Mahuad responded to the well-organized public pressure by trimming his gasoline price increase from 174 to 49 percent, but he left his banking policy in place. Although the opposition somewhat dissipated after what Mahuad deemed a major conciliatory move, CONAIE continued its call for a national dialogue on pressing public issues, which it referred to as a National Accord Forum. The Indians proposed to meet with the government to discuss ways to pull the nation out of its social and economic crisis. The request was backed by other popular organizations, notably the Coordinated Social Movements (CSM) that grouped together agricultural, oil, power, telecommunications, and other public employee unions under the leadership of Napoleón Saltos, a former socialist deputy who had helped plan Bucaram's ouster. Although Mahuad's rhetoric in response to the request for talks was positive, he never met with CONAIE or the CSM.

Doubts about the president's ability to handle public opposition to his policies as well as his aptitude for dealing with savage political infighting in Congress emerged as soon as he buckled over the issue of fuel prices in late March. Thereafter, predictions came with increased frequency that Mahuad would soon be removed. Emboldened by the general strike of the previous month and by the president's policy adjustments in the face of that shutdown, municipalities called for a day-long work stoppage in April. They demanded a greater share of the federal budget, insisting they were without resources to provide even basic services. Business leaders in Guayaquil proceeded to join the strike, and an umbrella group representing both labor unions and business organizations called for street marches and protests.

The April shutdown was called as Congress continued to grapple with Mahuad's plan to cut $400 million from the budget in 1999, thereby reducing the deficit, he hoped, from 7 percent to 3.5 percent of the gross national product. The package also restored the income tax and scrapped a number of tax exemptions. The IMF informed the president that its passage, along with reform of the financial system, was both urgent and essential to qualify the country for the three-year loan being negotiated. Mahuad was mindful of the fact that several of his predecessors, including Osvaldo Hurtado, León Febres Cordero, and Rodrigo Borja, had been denied IMF loans in the amounts requested for failing to adopt fiscal policies mandated by the fund. He hoped that an agreement by the end of May would enable him to renegotiate a significant part of the foreign debt, specifically the

$1.2 billion obligation to the Paris Club of creditor nations. The so-called good news was that with the IMF agreement, inflation would only hit 75 percent and the economy would only sink 6 percent. The bad news was that without the agreement, inflation would soar as high as 180 percent and output would plummet as much as 10 percent. But without a lean budget, IMF loans were highly unlikely. Nevertheless, the fractured and contentious Congress continued to debate rather than act on the proposal, as few had enthusiasm for an income tax and other measures to levy more and spend less.

Calls for the president to resign came with increased frequency. On April 21, Vice President Noboa was asked by an *El Universo* reporter about the possibility of Mahuad stepping down. He replied: *If President Jamil Mahuad should face some misfortune, and I hope that does not happen, or should he resign the presidency, and I hope that does not occur, I will assume the office with full constitutional powers.* [18]

In mid-1999 the president was cautioned about his performance by the military and the general public alike. On June 3 the head of the Joint Command, General Carlos Mendoza, warned Mahuad it was time Ecuador *changed course* in dealing with its worst crisis in a generation. Later in the month a forum was organized by former presidents Hurtado and Borja, along with prominent PSC member Heinz Moeller and other politicians, to discuss the role of the Armed Forces. The intent of the forum, according to General Mendoza, was *to maintain civil-military dialogue about fortifying democracy*. Mendoza later concluded the initiative had no impact on Mahuad, and the nation's politics continued to be dominated by *political party interests, electoral calculations, negotiating votes, all conditioned on individual economic interests, which put in serious danger Ecuador's fragile democracy.* [19] Later in the month, on June 29, Vice President Noboa reiterated to a *La Hora* reporter that he was ready to assume the presidency if called upon: *I am not a politician, I do not depend on any party line . . . I am not a person who can simply be resigned to what I see.* [20] Minister of Finance Ana Lucía Armijos defended the president and his administration by saying the economic crisis was not produced by Mahuad or by his predecessor. Rather, it had been developing for the last two decades and would not be resolved overnight.

Mahuad saw two immediate alternatives for putting the budget somewhere near balance: raising gasoline prices by lowering state subsidies, or creating new taxes. Once a satisfactory budget was passed, he said, the IMF would extend $1.4 billion in loans to move the country out of its economic crisis. To balance the budget for 2000, Mahuad needed $1.5 billion, which had not been raised in any form. He chose to increase gas prices, he explained, because that was *one of very few alternatives available to the president. If Congress gives me a rational tax system, gasoline prices would not have to rise.*[21] When the 7,000 taxi drivers who had shut down service in mid-March went on nationwide strike again from July 5–16 to oppose the fuel price increase in early July, polls revealed that they were backed by about 76 percent of the population. Calls for Mahuad's resignation multiplied. PSC leader and former president Febres Cordero declared on July 13: *The moment has arrived . . . if you cannot manage Ecuador, Mr. Mahuad, then go home before the people do what they had to do with Bucaram.*[22] The next day, former presidential candidate Ehlers expressed his opinion: *This is a crazy situation. One cannot govern with 90 percent of the people against you.*[23]

The taxi drivers were joined by thousands of Indians in a protest ignited by higher gasoline prices and plans to privatize state enterprises. CONAIE's Vargas announced his movement's support of the transportation workers and called for a moratorium on paying the foreign debt, an end to privatization of government enterprises, and administrative decentralization of the country. Thousands of indigenous people converged on Latacunga in Cotopaxi province, and a dozen protesters were injured when troops opened fire with tear gas on those blocking the bridge into the city. Radio and television stations were occupied in Tungurahua province, and within a few days Indians throughout the highlands clogged roads and kept food supplies and gas out of urban areas. Mahuad immediately mobilized the military, declared a state of emergency, and suspended freedom of association. Congress reacted by dealing Mahuad a heavy blow by voting to lift the state of emergency he had declared throughout the country.

On July 12, General Mendoza encouraged *reflection and dialogue between all sectors of Ecuadorean society in order to arrive at solutions, . . . the Armed Forces have the grave responsibility and duty of strengthening the state of law and being at the service of society, fundamentally with the poorest, and have responsibility for security and national defense*. His declarations, reported in *El Comercio*, urged that *politicians think in terms of national objectives, the private sector be honest, and unions responsible.* The military, he added, was suffering like everyone else *because we are part of the Ecuadorean nation*. Mendoza urged an end to *the notion that civil society is one thing and the military another . . . we have to think in terms of all of us forming Ecuadorean society*. Privately, the general concluded that to date Mahuad had failed: he *had completed one year of government without complying with a single one of his electoral promises.*[24]

The strike and demonstrations resulted in 578 arrests, three deaths, and dozens of injured. It ended on July 14 only when the president announced a freeze on gas prices at their previous levels until the end of the year and backed off from implementing an IMF recommendation to raise them 105 percent. Mahuad's comment about the twelve-day strike was: *I did not believe the reaction was going to be so great.*[25] Respected sociologist Simón Pachano, in an article published in *El Universo* on July 19, claimed the country had three alternatives: Mahuad's resignation, his assumption of dictatorial powers, or the continuation of a very weak government facing constant protests and damaging political accusations. Pachano concluded that *the only support for Ecuadorean democracy comes from the military, the North American government, and international financial organizations.*[26]

The president and the IMF failed to come to an agreement on refinancing old loans and advancing new ones. By late July, General Mendoza believed privately that *the IMF delayed agreements* because *the government and its economic advisers did not have sufficient credibility.*[27] On August 20 the public sent their leader a similar message in polls that showed what most suspected months earlier, namely, that the president's popularity had plunged from a high of 66 percent immediately after assuming office to a low of 16 percent. Mahuad had shrugged off his dwindling ratings as the price to be paid for making unpopular but necessary reforms. *Popular support,* he lamented, *is important, but it is a historical constant in Ecuador that a president arrives, has to make tough decisions, and falls out of popular favor.*[28]

August, September, and October witnessed a succession of financial disasters that culminated in the Brady bond default. The bonds were a type of debt created in the 1980s to help developing countries overwhelmed by their foreign financial obligations. They were named after U.S. Secretary of the Treasury Nicholas Brady, who led the effort to convert billions of dollars in existing Latin American loans into bonds backed by U.S. Treasury notes. Fearing that full repayment would not be forthcoming, many lenders followed the lead of Lloyds of London in 1995, when as part of debt restructuring with a number of nations it made a 45 percent writeoff on defaulted loans. The obligations were then turned into Brady bonds, and investors joined banks as major creditors to developing nations. Some observers hoped that the new arrangement meant fewer conflicts of interest between lenders and borrowers as bondholders, unlike bankers, were without business interests in debtor countries. In Ecuador, Brady bonds constituted about 36.3 percent of the foreign debt. Euro-bonds made up 3 percent, and World Bank, Inter-American Development Bank, and Andean Development Corporation loans 21.4 percent. Another 2.6 percent consisted of IMF credits, 11 percent came from transnational banks, 7 percent from the Paris Club, and the rest from other sources for the country's total liability of $13.6 billion.

In August, Mahuad informed the IMF that Ecuador would defer payment on part of the Brady debt. The country, he said, simply could not pay more than $52 million of the $98 million owed in monthly interest. His options were stark and dangerous: either miss the debt payments due in August, or let government workers go without their paychecks. In October he extended the moratorium to cover the entire array of external obligations. World markets were shocked, and the refusal to pay effectively put IMF negotiations on hold. Critics pointed out that the default was poorly handled; Mexico and Argentina had undergone similar problems, but with better planning they had gained a measure of assistance and cooperation from the United States and the IMF.

Congress, in the meantime, continued to debate more than act. Ecuador had requested $400 million from the IMF, which would lead to additional funding from other international lenders, but the lending agency's chief, Michael Camdessus, balked at the plea. He was determined that Ecuador perform major surgery on its economy to win loan approval, and that the country explore more options for internal sources of revenue. Without a budget and with the failure to win IMF loan approval, the president could not attract private investors to the largely inefficient oil, electricity, and telecommunications sectors. The partial sale of state enterprises, he hoped, would both bring in revenue and reduce state expenditures by involving private capital. But possible sales seemed stuck, as was everything else.

At the same time, international events took a turn for the worse. In September, Ecuador saw its first threat from a guerrilla force whose seriousness was difficult to evaluate but which nevertheless caused considerable concern. Part of the worry came from neighboring Colombia, with its long-standing and troublesome guerrilla movement and drug trade, and the fear that both might spill over into Ecuador.[29] Eight foreign oil workers and four tourists were kidnapped by an unknown armed group. Although the tourists were released within a few weeks, the oil workers were held for ninety-nine days. Rod Dunbar, one of them, said after his release that he did not want to discuss the $3.5-million ransom his employer, Alberta-based United Pipeline Systems, reportedly paid for his and the other workers' release, since it would only encourage more kidnapping. In mid-November, Ecuador's only oil pipeline was dynamited, which caused 36,000 barrels of crude to spill and the sucre to drop 4 percent. An organization calling itself the Popular Fighters Group (GCP) claimed responsibility and boldly painted its initials and slogans on walls throughout the capital. Guerrillas and terrorists added a new and potentially explosive dimension to the country's overall insecurity and discontent.

General Mendoza later wrote that throughout 1999 *the people waited for the executive to denounce the corrupt, privileged, and ambitious. But it did not happen*. Vice President Noboa urged Mahuad, for example, to order the arrest of one of the nation's most prominent bankers, Banco del Progreso head Fernando Aspiazu. The president's only response was, *What do you have against him?* On October 8, 1999, a political bombshell hit Mahuad. Aspiazu had finally been arrested by the Armed Forces on July 12 in Guayaquil. As the city was under martial law because of a crime wave, the military had the authority to apprehend him because he was already facing judicial proceedings. Mahuad neither ordered the arrest nor knew of it until it had taken place, for reasons that became clear in early October. General Mendoza later said that the minister of defense, General José Gallardo, *had done a slender service to the president. Without knowing it, and by his own initiative, he had arrested one of the major contributors in the presidential election campaign.*[30]

Like many bankers, Aspiazu had contributed heavily to Mahuad's electoral campaign and now felt betrayed by the president. During the campaign, Mahuad's vice presidential running mate had advised him, *Do not take money from Aspiazu or from other bankers. . . . For ethical reasons, you should not receive money from these bankers. Afterward they will pass the bill to you*. After three months in jail, Aspiazu divulged from behind bars that he had donated $3.1 million to Mahuad's 1998 presidential bid. He maintained that the money was accepted by the campaign's financial director, Mahuad's brother Eduardo, and that contrary to the Law of Political Parties it was never reported to the electoral tribunal. Aspiazu ended his denunciation of the president with the question, *Where is the money, Dr. Mahuad?* Mahuad admitted receiving over $30 million and confirmed that part of it was donated by various bankers whom he refused to name. The scandal broke when the nation was already seething over the freezing of bank accounts and over financial bailouts to the managers of those institutions. Thereafter, the assumption that Mahuad had corrupt ties to the banking community was the major catalyst in uniting disparate groups to oust him from office. General Mendoza later wrote that it became clear in October that the president *had shown sympathy, tolerance, and acceptance of that asked by those who had financed his electoral campaign.*[31] To make matters worse, in November the president refused to be questioned by a judge investigating claims that he and a handful of his associates kept millions from his 1998 election fund, including Aspiazu's multimillion dollar contribution. He claimed that Judge Alfredo Rijalba was without authority to issue summons to the head of state and that the groundless charges were politically motivated.[32]

After the revelations of October 1999, Mahuad was seen as just another corrupt politician who had enriched himself and shielded thieving bankers from prosecution. Former President Borja spoke for many people when he lamented that politicians should not be the watchdogs of banking and business interests but should defend the public interest. He labeled the bankers a group of gangsters.[33] Another former president, Carlos Julio Arosemena, said in an interview printed in *El Universo* on October 24 that *today is the worst moment in the history of the republic and President Jamil Mahuad is the worst president*. The newspaper *Expreso* editorialized that *President Mahuad no longer has the moral authority to lead the country*, and that in regard to Aspiazu and other corrupt bankers, Mahuad was *guilty for giving and for receiving*.[34] The Armed Forces shared these sentiments, including Colonel Lucio Gutiérrez, who thought corruption was beyond the ability of civilians to control. Earlier in 1999 he had publicly urged his superiors to devise a plan making the military the guardian of the nation's revenues. General Mendoza concluded privately about Mahuad that *nobody believed his words; they doubted his sincerity, and they noted a permanent incoherence between . . . what he said and what he did.*[35]

The Indian movement was equally indignant about corruption in general and the banking mess in particular. CONAIE's leaders claimed the bankers, with their own money, should resolve the problems they created, and they rejected the notion that money from taxes or monetary reserves should be used to bail out the private sector. They condemned the attitudes of Mahuad, the PSC, and the business community alike, calling them self-serving conservatives who favored giving public funds to wealthy, dishonest, and dangerous entrepreneurs while leaving Indian communities and the poor to languish. Mahuad had bailed out eighteen financial institutions at a cost of $1.2 billion while leaving millions of bank accounts frozen.

What Mahuad or others did with the $3-million Aspiazu contribution is not clear. Rodrigo Borja called the money in question *a donation without precedent in the history of the country.*[36] What is beyond dispute is that Mahuad was perceived as a leader who put his banker friends first, bailing out their institutions only to have many of them flee abroad with millions, while at the same time freezing the accounts of some four million people who were far less well off. His moral authority and ability to lead plunged, and he never recuperated.

In November, Mahuad spent much of his time in budget negotiations with Congress. His objectives remained a cut in spending and an austerity package to satisfy the IMF in order to obtain a $400-million loan. The lenders continued to insist that financial reforms and a tough budget be enacted before any new money was released. Ecuador had become a major battleground between the IMF and investors over the principles of lending to developing countries. The debate centered on whether creditors would forgive bond obligations, and, if so, who would pick up the tab. Some observers believed that a writeoff of the debt, in whole or in part, was the only realistic solution to the country's foreign debt problem, as interest payments consumed half of government revenues and the amount owed nearly equaled the gross national product. The country's foreign creditors looked to each other to forgive part of the debt. The Paris Club, a league of development and export-import banks from industrialized nations, was asked to renegotiate part of its $2.5 billion in Ecuadorean debt, but its members balked at the prospect until Mahuad first reached an agreement with the IMF. Any writeoff would have a huge impact on other larger developing countries who were also demanding restructuring of their far greater debts. With so much at stake, the IMF was determined to be sure that adequate financial reforms were in place and that all sources of available money from its own citizens were found before old debts were restructured and additional loans were made. In mid-December, U.S. Assistant Secretary of State for Latin America Peter Romero underscored the point when he visited Ecuador to make it clear that internal revamping and loans were inseparable.

A 2000 budget limiting the deficit to 2.5 percent of the gross domestic product was finally passed in November, but reform of the nation's financial system stalled. Proposed spending for 2000 included the equivalent of $22 million for health care, education, and other social welfare projects in Indian communities.[37] The projects were offered following negotiations with CONAIE that had begun in July. Mahuad was prepared to sign accords with CONAIE, but they were never formalized. CONAIE's Vargas thought the funds were insufficient and declared that the Indian movement *could not be bought with candy*.[38] Mahuad and his ministers believed that Vargas refused to sign off because he was running for reelection as CONAIE's head and feared he would lose if he were charged with accepting too little from the government. Indeed, Vargas was reelected on November 19, defeating the more radical Ricardo Ulcuango of ECUARUNARI.

With Plan Colombia, President Mahuad hoped to stem the growing tide of drugs, modernize an air base, and improve chances of U.S. support for international loans when in early November 1999 he struck a deal with the United States. He agreed to let the North American air force use Eloy Alfaro Military Air Base in Manta for ten years as part of a regional effort to reduce narcotic trafficking in Colombia and elsewhere in Latin America. Aside from the shared U.S. presence, the base would continue its usual operations. (Manta, a port community of about 170,000 where half the people depended on fishing, is 240 miles southwest of Quito on the coast in Manabí province.)

Unarmed U.S. planes from the base were to engage exclusively in antidrug operations. Surveillance planes would fly over Colombia to spot drug activity and radio information to Colombian police and military detachments, to abort the export of cocaine to foreign consumers. The major coca-growing area of Putumayo, on the border with Ecuador's northeastern Sucumbíos province, was a few minutes' flight time from Manta. The base was one of four negotiated by the United States to compensate for the loss of Howard Air Force Base in the Panama Canal Zone at the end of 1999. (Other bases were in Curacao and Aruba in the Caribbean, and in El Salvador.) After runways and hangars were revamped, two giant Super E-3 AWACS and 2 KC135 tankers to refuel them in midair, along with three small P-3 surveillance planes would replace the smaller Navy aircraft. These improvements would allow the United States to monitor air and marine activity far into the Caribbean on a twenty-four-hour per day basis and enable full resumption of the antidrug surveillance flights cut by two-thirds when U.S. forces left Panama.

As part of the deal, the United States pledged to invest a minimum of $70 million to modernize the base over the course of the pact. U.S. personnel were to offer community assistance in education, culture, and hygiene. Manta had been underdeveloped and violent by Ecuadorean standards during the 1980s, but it had become a modern city by the end of the 1990s. From 1997 to 2000 fishing grew 50 percent in value with increased exports to China and Japan, and a greater number of tourists came to safe, clean, and beautiful beaches. The U.S. military presence, backers believed, added to Manta's image of security. Moreover, while the rest of the country suffered a severe economic crisis in the late 1990s, the city underwent a boom.

The lease was closely linked to the so-called Plan Colombia, a $7.5-billion program announced in September 1999 by Washington and Bogotá to deal with the civil war fueled by the drug trade, a decades-old fight that had left 35,000 killed in the 1990s alone. The rebel Revolutionary Armed Forces of Colombia (FARC) dominated an area in the south about the size of Switzerland, and the government wanted to cut off its cocaine-based cash supply. The country supplied an estimated 90 percent of the world's cocaine and a large amount of heroin as well, and the plan sought to eliminate both. The Manta facility in Ecuador was a key part of the effort to cut Colombian drug production in half in six years. More than 50 percent of the country's coca was produced in Putumayo. As the province was largely controlled by FARC, it was impossible to untangle Colombia's civil war from the crackdown on coca growing and the drug trade.

Colombia's conflict was complex. In addition to the nation's armed forces, it included the 4,000-member National Liberation Army (ELN), formed in 1963 with its ideological homeland in Castro's Cuba; the 17,000-strong FARC, launched in 1966 as the military wing of the Colombian Communist Party; and a smaller rightwing paramilitary group, the United Self-Defense Forces of Colombia (AUC), created in the 1980s with ties to elements in the army to fight leftist guerrillas and narcotraffickers. During the course of the four-decades-long civil war, the central government lost control of over 40 percent of the nation's territory. The operations of the ELN, FARC, and AUC were funded in one way or another by the drug trade as well as by kidnapping for ransom. Most profits from the drug trade were made in the United States. The United States was the world's largest single market for cocaine, with an average of 300 tons of the white powder entering the country every year.

There is little indication that Mahuad thought the Manta pact would impact Ecuador's security, or that Colombian rebels, kidnappers, terrorists, coca growers, and innocent refugees would spill across the border as Plan Colombia progressed. To the extent that such matters were considered, Mahuad and his administration believed they would unfold whether the United States used the airstrip in Manta or not. Foreign Minister Benjamín Ortíz said the administration viewed cooperation with Washington as *a way of protecting the country from the problem of drug trafficking. We should realize that we have the world's largest criminal enterprise next door and that it can destroy us.*[39] Although initial protests against the accord were mild, nationalist voices were raised with increasing frequency against it; and as the months passed, the left and CONAIE were more outspoken in their opposition to the presence of U.S. forces on Ecuadorean soil. While many people supported the Manta base's use to fight drug trafficking, most opposed what they believed were simultaneous U.S. efforts launched from the airfield to put together an operation of the Colombian military to fight rebel forces. They argued that their neighbor's internal political struggle should not involve Ecuador, and a growing number feared the conflict would doubtless impact their country in multiple negative ways.

Increasingly blunt calls for Mahuad to leave office came in November and December. The nation's newspapers were bellowing, *The president is against the ropes, Mahuad cannot leave his labyrinth,* and *Mahuad evades questions*. In a November 15 interview with a reporter from *Expreso,* Vice President Noboa for the first time spoke out: *The truth is, that something the public will never pardon this government for is the bank holiday and the freezing of deposits . . . I never knew it was going to happen, and when I found out through the media, it had a tremendous impact on me*. Regarding Aspiazu's campaign contribution to Mahuad, he declared: *I know nothing of these funds; those directly responsible are Mr. Yulee and Mr. Eduardo Mahuad. You have to ask them the questions and they are the ones who have to give all the answers*. Noboa believed Mahuad was not guilty of stealing the banker's contribution, but when asked if he would put his hands in the fire for the president, he replied, *I only put my hands in the fire for Gustavo Noboa.*[40] Seven days later, Rodrigo Paz was also interviewed by an *Expreso* reporter and gave the following reasons for wanting Mahuad to resign immediately:

> *The bank holiday was done to help Mr. Aspiazu. . . . It is not good to generalize, but he (Mahuad) has been the hostage of certain bankers who helped him profusely, but one must ask why they gave such large amounts. Because they are good Ecuadoreans? Or because . . . they wanted to be favorably treated in these difficult times? Mahuad fell in the hands of these people and could not distinguish between the ethical and unethical, between the moral and the immoral. . . . Do you believe that Mahuad did not know of the crises of some of the banks, among them the Progreso? I was the first to warn that Mahuad would not properly manage the country, that he was without ethics, and many criticized me when I said it. They accused me of being envious, but now the country feels that Ecuador is like a business without a capable director. Vice President Noboa is a fresh man, an academic, a man whose honor nobody has questioned. I believe that as president he would surround himself with capable people and enter into a dialogue with politicians with different attitudes*. On December 29, Paz added that *the time will arrive when the Ecuadorean Army will have to make a*

decision about the economic and institutional crisis, given that the military does not want to enter the streets to repress their Ecuadorean brothers who protest against the precarious conditions of life and the bad government of Jamil Mahuad.[41]

Minister of Defense Gallardo attempted to defend the president but was ridiculed when on November 18 the best he could muster was *he who is without sin should cast the first stone.*[42] A broad range of groups demanded that Mahuad resign, among them CONAIE and other Indian organizations, farmers, labor unions—including the powerful oil and electrical workers—businessmen, students, and a wide array of politicians. All called for a continuation of antigovernment demonstrations throughout the country. While considerable publicity was given to 1,000 peasants who walked or rode their horses from the northern Amazon region to Quito on November 21, theirs was only one of what seemed like a never-ending series of protests.[43]

Speeches at demonstrations routinely blamed the president for poverty, high unemployment, spiraling inflation, and corruption. Pressure for Mahuad to resign crested in early December. The Chamber of Small Industries of Guayaquil made the call, as did the Federation of Chambers of Agriculture of Ecuador, labor unions, social movements, and Indian organizations. CONAIE planned a massive demonstration to be launched on January 15, 2000, called the Levantamiento Indígena, to oust Mahuad. The Armed Forces responded rapidly with a Joint Command press bulletin of December 24, which made it clear they would not permit the country to be destabilized. The top brass urged all Ecuadoreans, CONAIE included, to reach a consensus to overcome the country's worsening social and economic crisis.

At year's end, discontent with Mahuad came hard and fast from every direction. Led by the chief of the Joint Command, General Mendoza, military leaders warned the president that current conditions simply could not continue. According to Mendoza,

On December 27 we gave the president an analysis of where the nation was at that juncture. We told him that he projected a weak image, had fragile support in Congress, faced multiple social, political, and economic demands and antagonism from various fronts, and that his power had been reduced and his capacity to unite the people had been limited. We told him that the army was in a central position, with forces converging on all sides. Against him were political parties, social movements, chambers of commerce and banks, transportation workers, unions, Indians, the media, the agricultural sector, students, professors, and Congress, and even the international organizations lacked enough confidence in him to make loans. He had everything against him.

Mendoza related that the generals and admirals handed the chief executive a list of suggestions that included running a more open government, arresting crooked bankers and addressing corruption, changing monetary authorities, ending currency speculation, raising salaries, unfreezing bank accounts, and opening dialogue with social movements and groups in Congress. The general told Mahuad that *the elements which make up the structure of the state are under threat, to the point where the possibility of its survival is under question. . . . It is very difficult, as an Ecuadorean and military man, to accept this reality for our country, and even worse that the severity and the infinite number of threats keep us from being able to define which one is the most dangerous and most imminent.*[44] Mendoza then presented four scenarios to the president: changing government policy immediately, resigning and being replaced by Vice President Noboa, having someone other than Noboa replace him, or convoking a national constituent assembly to make needed economic and political reforms.

The senior military officials could not have been more frank. But as bad as matters were, they soon grew worse. As the new year opened, Mahuad concluded that 1999 was *one of the worst years in Ecuador's history.*[45] Calls for his resignation came daily. Three former presidents—Borja of ID, Febres Cordero of the PSC, and Hurtado of DP—all asked him to relinquish his office. Borja declared that Mahuad

> *is a theorizer, more or less coherent with simple things, but with an almost pathological incapacity to get things done . . . those that suffer this sickness have the talent to theorize about the human or divine, but make mistakes when they have to act. . . . We have to think of a constitutional change. Incompetence, the intrigues of economic interests, the lack of capacity for work, the lack of leadership, will not serve to move this country ahead. . . . What we want is a succession to see if Noboa is able to do what Mahuad could not during all these months; he is the person the constitution establishes and the people have elected to succeed in this situation.*[46]

A private poll conducted during the first week of the year reported that 91 percent of Ecuadoreans disapproved of his handling of national affairs and that 53 percent wanted him to step down. Mahuad was briefly buoyed when U.S. President Bill Clinton called on January 7, 2000, to praise him *for his commitment to constitutional order and to implementation of painful but needed economic reforms.*[47] Although he had already made the decision on January 4, Mahuad did not mention a measure he hoped would save his presidency: namely, his plan to substitute dollars for sucres as the country's currency. He presented the proposal in a televised address two days later.

On January 8, 2000, a group of high-level military officials, including Defense Minister Gallardo, held an emergency meeting; afterward they met with Mahuad. The generals and admirals knew that in a few days the president would address the nation on television to announce his plan to substitute dollars for sucres. They implored him to also speak on other vital issues to save his sinking administration and to restore confidence in his leadership. Specifically, the high command asked the president to extradite corrupt bankers and freeze all of their assets, offer a realistic plan to end corruption at the customs houses in Guayaquil and elsewhere in government, and present adequate measures to modernize the country. The defense minister was particularly preoccupied with the corruption issue, having been recently humiliated in a series of newspaper attacks. The most vicious came from Francisco Huerta, assistant director of *Diario Expreso*, who wrote that Gallardo not only defended an inept president but a hopelessly corrupt one as well. Opinion polls showed that most Ecuadoreans shared Huerta's assessment. The January 8 meeting between the military and Mahuad was only one of many that had taken place throughout 1999. Some officers in the Armed Forces saw the conferences as attempts to help Mahuad; others, particularly in the cabinet, thought they constituted inappropriate military pressure.

General Mendoza informed Mahuad that the Armed Forces saw the following differences between him and Vice President Noboa: *You, Mr. President, are in the center of the eye of the hurricane; you cannot form a national consensus, Dr. Noboa can . . . Dr. Noboa has time to call a popular referendum . . . Dr. Noboa is not accused of corruption . . . Dr. Noboa is not blamed for the banking holiday and the freezing of funds. . . . They are not claiming that Dr. Noboa received $3.1 million and are not demanding from him the return of that money.*[48]

By early January 2000, the head of the War Academy, Colonel Jorge Luis Brito, had come to the same conclusion: *the vice president had distanced himself from Mahuad in the previous weeks*. Brito later added that the vice president *at the same time had met with many in his residence in Punta Blanca, especially in the final days of December 1999. Generals Mendoza and Sandoval had been invited to those meetings.*[49]

Ecuador approached hyperinflation in early January 2000, as the sucre finished the first week of the year down 17 percent in seven days after diving a disastrous 67 percent the previous year. Financial markets, hoping the Central Bank would prop up the currency, were disappointed when the government announced on January 5 that it would continue the nearly year-old system of "free-floating" exchange in which the open market would determine the sucre's value in relation to other currencies. Among other things, the decision meant the state could print all the money it needed to avoid making difficult and politically explosive financial sacrifices.

Currency debates were hardly new in Ecuador or elsewhere in Latin America. The "fixed" versus the "flexible" currency debate usually presented the choice between the state setting a fixed exchange rate for currency, which could be realistic, undervalued, or overvalued in relation to other countries' currencies, or taking a flexible approach and letting the money "float" and find its value in the free market. (A recent example of the dire consequences of an overvalued currency was Thailand's in 1997. Thai financial houses had borrowed heavily in dollars and had lent them to Thai businesses. When the government failed to keep the baht at a fixed rate against the dollar, the currency crashed 30 percent and virtually all of the country's major financial houses failed. Many businesses did as well, unemployment rose, and the Asian financial panic ensued, with a worldwide impact.)

So-called dollarization is the most radical form of fixed currency, but there are important differences between that approach and "fixing" an exchange rate. By adopting the U.S. dollar, Ecuador would give up the ability to print its own currency and thereby lose the revenue it could gain from creating more money, regardless of the impact on inflation, and would also lose the power to adjust exchange rates to absorb external shocks to the economy such as drops in the value of oil or bananas. If prices plunged for export products in world markets, the government would face domestic pressure for either a currency depreciation to make prices appear lower or for a cut in costs. Both would result in a drop in real wages, which is why labor unions opposed fixed currency policies as well as dollarization. On the other hand, when faced with inflation, the wealthy could convert their sucre incomes to far more stable U.S. dollars to protect themselves. They traditionally sent much of their capital to the United States and other safe havens abroad.

Negotiations in early January with the IMF on securing new credit were fragile at best, with the economy's accelerated downward plunge and with inflation soaring to ever-greater heights. For months, Mahuad confronted mounting criticism for putting off crucial decisions that might halt the slide. But dramatic and unexpected action finally came on January 9, when the president revealed in a televised address that Ecuador would abandon the 128-year-old sucre and become the first South American nation to adopt the dollar. Mahuad, through Minister of Defense Gallardo, asked the military high command to appear with him on television, but General Mendoza refused on the basis that the military should be independent of policy making. Most analysts thought the decision to scrap the sucre was taken hastily and unilaterally and was motivated more by politics than economics. Mahuad's announcement took virtually everyone by surprise, including international lending agencies and the U.S. government. The president said he would send his plan to Congress for 30-day emergency action, and it would automatically become law if the legislature did not take action during that time. Mahuad insisted it would curb inflation, bring down interest rates, and spur investment. Above all, the currency switch would prevent politicians from printing more money, since only the U.S. Treasury had that authority. He apologized for his delay in trying to stabilize the economy; and to dramatize the importance of his action, he divulged that his entire 15-member cabinet had been asked to resign to give him greater maneuverability in addressing the crisis. He accepted only the resignations of two officials who had criticized him for failing to focus on corruption, the banking crisis, and a host of other critical issues that threatened stability.

The predictable chorus of support for currency conversion came rapidly from administration members. *This is going to return confidence to the country, stop speculation over the exchange rate, and stop the rapid increase in prices,* Finance Minister Alfredo Arízaga assured the nation at a press conference. *There will be a progressive process in which the sucre will be pulled out of circulation and dollars will become the currency.* [50] Juan Falconi, the minister of production, and future banking superintendent predicted that *the Central Bank will stop issuing the local currency and there will be no more devaluation, and inflation will decrease from the tremendous levels we've had in the last year to international levels.* [51] Mahuad's team claimed the currency switch would work because Ecuador had enough dollars to implement it. The Central Bank issued assurances that its liquid dollar reserve was more than enough to cover the sucres in circulation. Since about 60 percent of the country's bank accounts were already in dollars, supporters argued the conversion would not be traumatic for the country. The U.S. Federal Reserve estimated that about $580 billion were in use worldwide, two-thirds of it outside of the United States.

The effort to end the economic nosedive cheered the nation's currency and bond markets, and both improved slightly on January 11. A poll taken on January 10 by Market, a private research firm, showed Mahuad's popularity more than doubled to 22 percent after he broadcast his dollar design, a significant rise from his previous month's 9 percent approval rating. The sampling also showed 59 percent of the public in favor and 40 percent opposed to giving up the sucre.

Yet many people were sanguine about the switch. After months of ongoing crisis and repeated promises to stabilize the currency, a number of prominent analysts and investors expressed skepticism that the plan would revive the economy and wondered if it could be implemented. Pablo Better, president of the Central Bank, resigned in opposition on January 10. Better left before the bank's four remaining directors voted in favor of the proposal, doubtless influenced by Mahuad's threat to ask Congress to fire them.

Repeated failures to address the economy's drift toward collapse had already destroyed faith in the president's ability. His dollar plan struck many observers as a desperate, ill-conceived, and futile attempt to prolong his doomed administration. An unexpected impact of making the proposal when he did, moreover, was to animate further groups already determined to bring him down, including CONAIE and the labor movement. While the daring move was meant to calm economic anxiety by guaranteeing some measure of stability and shoring up support among business leaders, it nonetheless had a radically contrary effect on the demonstrators who, weeks earlier, had decided to pour into the capital in mid-January.

CONAIE and labor leaders claimed conversion to dollars would devastate the nation's seven million poor people, particularly the Indians, and would raise prices while salaries remained low. They vowed to fight. *The dollar may be fine for Mestizos and the big folks, but we are peasants and do not know how to manage dollars,* worried Apolinario Quishpe, a 51-year-old farmer who made a 55-mile protest march to Quito from his home in Latacunga. *Many of us do not know how to read and write and do not understand what this is about. We feel cheated.* Augusto Aguirre of the Guayas Transport Workers Federation branded dollarization a *medicine worse than the illness,* and 13,000 of his federation's drivers in Guayaquil blocked the main streets of the city as part of the nationwide movement against Mahuad.[52]

Economists and others divided over the wisdom of adopting dollars. Some said quick monetary fixes were no substitute for overhauling a weak banking system and for improving fiscal policies. The head of the New York-based Latin American brokerage firm BCP Securities, Walter Molano, cautioned that *a country that dollarizes or moves to convertibility needs fiscal discipline, and that is Ecuador's biggest problem. You need a firm commitment to counter cyclical fiscal policy, with large sur-*

pluses when you are in recovery, and to things like labor reform that Argentina still needs . . . but survival is the issue and it could give them some respite.[53] Ramiro Crespo, president of Analytica Securities, a Quito broker-dealer, concluded that *Dollarization is like trying to lose weight by wiring your jaw shut. It is a desperate measure.* The official U.S. response was cool. Treasury Secretary Larry Summers commented that adopting the dollar alone would not be a quick fix for Ecuador; fundamental financial reforms would still be needed.

National sovereignty and pride entered the debate. León Roldós, socialist legislator and rector of the University of Guayaquil, cautioned: *More than a dollarization of the economy, this measure can be considered the Panamaization of the economy.* Panama's former finance minister, Guillermo Chapman, added that with no currency or central bank of its own, dollarized Panama had *no way . . . to adjust the economy to external shocks in the short term. The key is not just dollarization but full integration of the economy in the international financial system.* Architect Arcesio Vega was indignant: *People are familiar with the sucre. Within it is narrated part of our history. Now we have these coins with faces that are completely unknown to us, distant from our history, from what we are.* The national press was filled with articles that mourned the second death of the sucre.

Citigroup cochairman John S. Reed countered that while countries embracing the dollar might initially lose some flexibility in formulating domestic policy, its adoption would eliminate devaluation and result in lower interest rates. It might even reduce income inequality, Reed reasoned, since most workers were paid in sucres that lost value immediately, while highly paid businessmen were often compensated in far more stable dollars. Dollars for everyone would put labor on a par with executives in terms of stable pay.[54]

Still others saw conversion as inevitable, not only in Ecuador but also throughout Latin America. *I think it will all be dollarized,* predicted Steve Hanke, professor of applied economics at Johns Hopkins University in Baltimore. *There's a 90 percent chance that all of Eastern and Central Europe and the Balkans will be on the euro by the end of 2002, and by 2005 most of Latin America will be dollarized.*[55] The success of Europe's common currency in lowering inflation and stimulating rapid growth in Spain and Portugal, he argued, would make a single currency for Latin America both plausible and attractive.

Clearly no consensus existed among professional economists, financial analysts, social movement leaders, politicians, or others regarding the plan's likely effect or viability. Understandably, no agreement emerged among the population at large. Far more important in the short term was the fact that Mahuad's *foot on the accelerator*, as he called his dollar plan, unintentionally spurred thousands of demonstrators to speed up their efforts against him.[56] To many, the president seemed so desperate that he would try anything to survive politically, even if it meant a radical change in his own views. As late as December 27, 1999, opponents pointed out, Mahuad had told newspaper editors and others that he was against substituting dollars for sucres. He cautioned at the time that the scheme was too risky, would severely damage negotiations with the IMF, and would more deeply divide already fragmented public opinion. On the later score, at least, he was right. Currency has traditionally been a symbol of power and identity, and many already disgruntled Ecuadoreans feared a good measure of both was being lost.

A consensus failed to form on the conversion rate. Mahuad proposed 25,000 sucres to the dollar, which would leave the country with about $1 billion with which to pay the foreign debt. Others believed a more realistic conversion rate was between 14,000 and 18,000, which would make sucre holders between 44 and 28 percent better off than they would be with the 25,000-to-1 proposal. At the same time, businessmen who owned the transportation sector pressured Mahuad to guarantee that the payment of their debts was owed in dollars, and that they pay back the government at the sucre-to-the-dollar rate that existed when the dollar debts were acquired. (Most debts were acquired when the sucre stood at 5,000 to the dollar.) Mahuad refused their demand, which, if accepted, would have benefited the businessmen fivefold.

On January 10, 2000, the Central Bank approved the dollar plan, and Mahuad was expected to send it to Congress on January 21 along with legislation to open more oil, electricity, and telecommunications to private investment. General Gallardo was upset with Mahuad's televised address and told him so in no uncertain terms. At a cabinet meeting the defense minister vigorously reiterated the military's request of January 7; further, he urged Mahuad to address the nation on issues other than the national currency and to show bold leadership in dealing with corruption and other vital matters. Gallardo's strong words were backed by the agriculture minister, Salomón Larrea, but the president was too closely associated with the unpopular bankers to probe their scandals in depth, a fact that doubtless conditioned his weak references to the explosive subject. His response to the January 11 urging of Gallardo and Larrea was to approve their resignations. The corruption issue was at the heart of Mahuad's decision to remove Gallardo. Colonel Brito believed that *his departure came, among other motives, because of the government's displeasure with his detention* (in July 1999) *of the banker Aspiazu, who later* (October 1999) *involved Mahuad with the matter of funds to finance his campaign, which was one of the things that detonated his calamity.*[57]

Mahuad replaced the defense minister with Joint Command chief Carlos Mendoza, a man who many believed resented Mahuad for passing him over earlier in the administration for Gallardo. At the time of his selection, Gallardo was on inactive service. During the recently ended hostilities with Peru, Gallardo and General Paco Moncayo had gained considerable popular fame and respect. Unlike Gallardo, Mendoza was on active duty and thought he was in line for the post of defense minister. As 1999 ended and 2000 opened, rumors were rampant that Mendoza was undermining Gallardo and that the president was influenced by those efforts; he had become unable to distinguish between men such as Gallardo, who backed him while offering constructive criticism, and others such as Mendoza, who were less critical but sought to sack him. A cavalry school graduate, General Mendoza's 38-year military career included stints as a helicopter and fixed-wing aircraft pilot and training in the United States at Lakeland Air Force Base in Florida and at Fort Rucker, Alabama. After January 11 he served as both defense minister and head of the Joint Command.[58]

As these top military changes occurred, Mahuad announced in an address to Congress on January 15 that the government did not have sufficient money to release the bank accounts that had been frozen in March 1999. The announcement was received with harsh criticism by most deputies and the public. The president's words came when thousands of Indians entered Quito to participate in the January 15 Levantamiento. Their numbers grew daily, and on January 17 police teargassed 700 demonstrators in the streets of the capital. Two days later, CONAIE's leaders met with the Joint Command in a conference that had the president's approval. The fact that the meeting took place at all was a clear sign of the seriousness of the administration's deteriorating situation. The movement's directors told the senior officers that they rejected Mahuad and his government. They wanted the nation to acknowledge the legitimacy of *a government of national salvation*—namely, CONAIE's Parliament of the People, which had just convened in Quito.[59] It was not the first meeting between the Joint Command and CONAIE, but for some in the military, including Colonel Lucio Gutiérrez and other officers, the conferences encouraged the belief that the top command was signaling a change in government. Colonel Brito drew an important conclusion from a written communiqué from the Joint Command issued earlier on January 5. Instruction No. 01-20 read: *Permit the Indian and social movements to have their protests in an orderly way, giving priority to passive methods, avoiding violent confrontations that could make the situation worse.* Brito reckoned, as did others, *that the Armed Forces had decided not to employ arms against the people.*[60]

On January 20, as upwards of 10,000 Indians, peasants, students, and workers marched through the streets of Quito, General Mendoza announced that the military would defend the constitution and that the Indians and others should press their claims through the democratic process. But one day later, on January 21, thousands of Indians occupied Congress with the aid of dozens of sympathetic junior officers and soldiers. The Junta of National Salvation was declared, and later in the day General Mendoza demanded that Mahuad step down.

BARRIERS AND DIVIDES

Early evening, January 21, 2000. At the presidential palace, policemen and soldiers are on one side of the barbed wire, Junta of National Salvation supporters on the other.

The Indian uprising of January 15 was the third major one of Mahuad's administration; the first and the second had ended with gains for the Indians, but not nearly enough to satisfy them. Gustavo Noboa later said about Mahuad and the Indian movement: *the second Indian uprising should not have had any reason for being. Mahuad did not comply with any of the offers he made during the first protest, nor with offers he made afterward.*[61] Few took CONAIE seriously, including Mahuad, who thought that he could keep order by declaring a state of emergency and bringing out the troops. It had all happened before, most recently in March and July 1999, when CONAIE conducted massive demonstrations demanding that the president give up. As early as 1991 its leader, Luis Macas, had led huge protests in Quito, entered Congress with many of his followers, and threatened to form a rival government if the Indian movement's demands went unmet. On these and other occasions, the government survived.

As the year 2000 opened, Ecuadoreans continued in the worst depression since the 1930s. Recession, inflation, unemployment, and poverty, now chronic, rocked political life from top to bottom. Former president Febres Cordero bleakly described the nation's condition: *Internal confidence is destroyed. The country's image is horrible. If we don't create jobs, if we don't generate production, if we don't stabilize the country's economy, Ecuador is finished.*[62] Rigoberto Villarreal, a 34-year-old Quito taxi driver, offered his grim assessment: *The ruling class has destroyed the country. We've got to kill the entire political class and begin anew.*[63] Deposed president Abdalá Bucaram, from exile in Panama, saw nothing but deterioration: *The increase in inflation and high interest rates, the fall in monetary reserves, the growth of the debt, all demonstrate the absolute failure of the government.*[64] Similar sentiments prevailed in every segment of society. Many Ecuadoreans simply left the country. From 1995 to 2000, between 8 percent (one million) and 16 percent (two million) of the population departed for Spain, Miami, New York, and elsewhere; from 1997 to 2000 it is estimated that the number of Ecuadoreans living in Spain rose from 3,000 to 70,000; about 44 percent of them worked as domestics. Azuay province in the southern highlands was particularly hardhit by the flight of males.[65]

The success of Ecuador's attempts at democracy, like those of many nations, often depended on the state of its economy. The economic engine inherited by Mahuad was in dire straits, and it grew progressively worse until he was deposed. Growth was a mere 0.4 percent in 1998, and inflation ran at 43.4 percent. On both scores it was Latin America's most dismal performance, but it deteriorated even more in 1999, as output dove 7.5 percent and inflation topped 60 percent. Ecuador's sucre fell from 7,000 to 25,000 to the U.S. dollar. This plunge was unprecedented. In 1920 the national currency had stood at 2.07 to the dollar, in 1930 at 2.10, in 1940 at 5.0, in 1950 at 15.6, in 1972 at 25, and in 1981 at 33. But in 1984 it was at 118, in 1988 at 499, in 1992 at 1,874, and in 1996 at 3,500. When Mahuad took office in 1998, one dollar brought 6,500 sucres; in 1999, 20,100; and in 2000, 25,000. As the year 2000 began, most economists predicted that without bold corrective measures the currency would soon slip into hyperinflation, as in four days alone in early January it lost 28.5 percent of its worth. Throughout 1999 the price of gas, electricity, cooking and heating fuel, flour, rice, sugar, milk, and other staples soared, and the cost of drinking water catapulted 120 percent by year's end. The year 2000 promised more of the same, and during the first month inflation reached its highest monthly rate in thirty-two years, climbing to 14.3 percent. Strikes, protests, and antigovernment demonstrations increased in tandem with the surging inflation and soaring prices of necessities.

Virtually all sectors of society suffered. The Central Bank reported that commerce shrank 12.1 percent in 1999, followed by a decline in construction of 8.9 percent and in manufacturing of 7.1 percent. Only agriculture expanded, at a modest 1.8 percent. *We are in the dumps*, summarized Gustavo Pinto, president of the Pichincha Chamber of Industries.[66] The effects were felt in employment, education, and health care. Only one in three persons in the workforce had a full-time job, and the official unemployment level was 15.8 percent. Subemployment, defined as those who work less than forty hours per week against their wishes or receive less than the minimum wage though working full time, was 56 percent in 1999 and rising; it reached 65 percent in 2000. The minimum wage in the country, which had been raised during Mahuad's tenure, was still only $96.64 per month at the end of 2000; and, according to the United Nations, more than 70 percent of the population of 12.4 million lived in poverty.

About 750,000 children who should have been in school were not, as most worked to help support themselves and their parents. About 10 percent of the families in the highlands and Oriente took at least one of their children out of school for financial reasons, a situation which in late December 1999 led President Mahuad to announce a $5 monthly education credit per student. Despite the credit, spending on education had fallen precipitously since 1981. The fall was gradual: from 5.5 percent in 1981 to 5 percent in 1987, to 2.8 percent in 1992, and to .07 percent in 1999. And in 1999 alone, the number of people using public hospitals in Quito fell 40 percent; the sick simply could not afford even the minimal fees for service. In 1981 spending on public health took up 10 percent of the budget, but in 1999 it consumed less than 2 percent. Meanwhile, payment on the foreign debt absorbed a sum three times as large as that spent on social services and public works. Expenditures on education, health care, and social welfare were projected to continue to shrink. In October 1999 they stood at 19.5 percent of spending for the year, and the president projected a fall to 17.1 percent for 2000, a reduction of about $76 million. Over half of the budget went to pay the debt. UNICEF experts calculated that no country should dedicate more than 20 percent of its budget to debt, if it wishes to make social investments.[67]

The poverty rate in 1999 stood at 62.5 percent of the population, up from 46 percent in 1998. UNICEF reported that between 40 and 50 percent of children under age five were malnourished. The poverty rate was high for the Latin American region, which the United Nations' Economic Commission for Latin America and the Caribbean found to be 36 percent in 2000, up a point from the 35 percent level of the 1980s. Ecuador, moreover, had slipped from number 72 to 91 in the United Nations' Human Development index, which ranked 174 countries. As dismal as these and other official figures were, many observers thought they missed the mark and were set too low. There were more undernourished children under age five in Ecuador in 1998 than anywhere else in Latin America: 17 percent, compared to 11 percent in Peru and 14 percent in Mexico. Crime was up as well, as the number of homicides climbed 43.7 percent between 1900 and 1999; most attributed the rise to unemployment, lack of opportunity, and poverty.

Wealth continued its uneven distribution and did so along ethnic lines, with the indigenous and black populations bearing a disproportionate share of the poverty. They suffered the most from the soaring cost of necessities, as they had rarely been able to purchase more much than the basics. In Chota, a community in Imbabura province inhabited almost entirely by blacks, the majority earned $1.50 per day and worked a minimum of six hours. When asked what any government had done for the town in the last thirty years, most responded "nothing," aside from a bridge built during the Hurtado-Roldós era. Jim Brown of Chota explained: *We are too poor to protest—we cannot afford to lose a day's wages to do something like that.*[68] Per capita income in 1999 was about $1,150 per person, but 24.5 percent earned around $44 per month or $528 per year, about 40 percent of that of 1994. The richest 20 percent of the population owned 73 percent of the nation's wealth, and the top 3 percent of them absorbed 44 percent of what was produced. At the other end of the spectrum, the proportion of those living in poverty in rural areas was set at around 90 percent, with 60 percent of them surviving in conditions of near destitution. *The situation is truly tragic,* lamented Bishop Raúl López of Latacunga. *The government has cut the budget for everything in the social sector in order to satisfy the demands of the International Monetary Fund and to make payments on our foreign debt.*[69]

With a public sector debt load of about $13.7 billion, which nearly matched the gross national product of $14.5 billion, Ecuador roiled markets worldwide in September 1999 when it defaulted on a large share of its international loans. It became the first nation to default on its Brady bonds. Ecuador's failure to pay on its loans discouraged prospective lenders and investors from making new advances and left the country in a serious financial and political state. Yet annual payments on the foreign debt cost the country three times more than it spent on social services and public works. A full 54 percent of the budget was devoted to interest payments in mid-1999: 23.3 percent to pay interest on the foreign debt, and another 30.6 percent on the nation's internal debt.

The economic contraction had severe political consequences, and they intensified as projected financial figures grew grimmer toward the end of 1999. In November the Central Bank expected the economy to shrink 6 percent for the year, but only a month later it darkened the already bleak forecast to a 7.3 percent plunge. Similarly, while inflation for the year reached 60 percent, from November 1999 until Mahuad was deposed less than three months later, it soared a full 53 percent. And it got worse. A 16-year record was set in January 2000, when in that month alone inflation topped 14.3 percent. If the pace continued, it would translate into a 172 percent annual loss in the value of the country's currency.

Around half of Ecuador's revenues came from oil, its largest export and its main means of paying back a huge foreign debt. Oil prices collapsed yet again in 1998, and although they made a strong threefold recovery in 1999 and continued to climb to $24 per barrel in early 2000, the added revenue was not enough to expand the contracting economy or significantly improve overall financial conditions. For the past two years, moreover, the price of bananas had dropped, and the devastation caused by the El Niño required massive emergency spending. As noted earlier, in 1997–98 the weather pattern-generated storms, floods, and landslides left 20,000 homeless and severely damaged over 1,300 miles of roads. The losses were calculated at $2.6 billion.

In sum, slumping revenue, a contracting economy, high inflation, pressures to spend from a host of well-organized groups, and demands of international lenders that the country practice greater austerity before old loans were renegotiated and new ones made—all of these pressures intersected to thwart passage of a budget. Mahuad's belt-tightening program aimed at balancing the budget by 2002. He hoped economic discipline would unfreeze a loan from the IMF, which in turn would increase investor confidence and allow the nation to borrow larger sums from multilateral lending agencies. In January 2000 the lender was still awaiting a tougher 2000 budget and banking reforms before approving new loans and restructuring old ones. Congress was splintered; no party held a majority, and coalitions were as difficult to build as they were fragile. Demands mounted for Mahuad to take decisive action to meet the economic calamity, but his DP held only thirty-six of the 123 seats in Congress, and reaching agreement with the opposition proved illusive. Few deputies favored raising taxes and hoped instead that petroleum prices would continue to climb.

Not all of Mahuad's difficulties were rooted in the nation's financial performance. Yet hard economic times usually shorten political lives, and what he termed Ecuador's *worst crisis in seventy years* was assuredly a major factor in trimming his own.[70] His experience bore witness to the statement of José María Velasco Ibarra, Ecuador's recordholding five-term president, who reiterated throughout his career that *Ecuador is a very difficult country to govern*. Few disagree. During the 170 years from the birth of the republic in 1830 until 2000, the country promulgated nineteen constitutions, or one every 8.9 years. Over the same span of years, the nation had 111 executives—presidents, interim presidents, and governing juntas—counted by separate investitures of power. Ecuador averaged one newly invested government every 1.5 years.

7

Levantamiento Indígena

Ecuador's presidents had come to expect demonstrations and strikes against the austerity measures they deemed necessary to revive the economy. Over the course of 1999, Jamil Mahuad became particularly well accustomed to them. In March, after hikes in the value-added tax, and gasoline prices and the decision to freeze bank accounts, indigenous groups and transportation workers had joined forces to disrupt Quito for nearly two weeks. A series of similar demonstrations and shutdowns proceeded throughout the year. Antonio Vargas of CONAIE served as the prime mover in virtually all of them. In January 2000 he again assumed a central role in a broadly based movement to paralyze the country, force Mahuad to step down, and, according to CONAIE's plans at least, replace the entire structure of government with a new one.

Just before the massive protest, the so-called Levantamiento Indígena, the Joint Command of the Armed Forces issued a statement on January 8 rejecting *any attempt to break the legal order* and calling for a solution *within the constitutional and democratic framework*. It was similar to their December 24, 1999, release that had warned demonstrators and had assured the government that stability would be maintained. Although Defense Minister José Gallardo told reporters the military *fully supported the government*, the Joint Command's statement pointedly did not mention Mahuad by name.[1] The Armed Forces had backed the president in past crises; many saw the January 8 statement as a signal that they now offered only lukewarm support.

When the president presented his proposal to substitute the dollar for the sucre, he hoped to take the steam out of the movement scheduled to unfold on January 15. Following almost daily attacks for failing to revive the ailing economy, his dollar proposal seemed to have taken the initiative away from his opponents. Polls showed a majority of people believed the proposal worth implementing, but they did not reflect the feelings of thousands of CONAIE members about to march on the capital. The Indian movement stood adamantly against giving up the sucre. It called for a moratorium on paying the foreign debt; an end to efforts to privatize strategic economic enterprises, including oil, gas, and electricity; the detention of corrupt bankers and an end to corruption; and a radical change in economic policy.

As the date for the Levantamiento approached and thousands of indigenous people prepared to enter the capital, CONAIE's Parliament of the People convened in Quito on January 11. Delegates recently chosen by provincial assemblies met in the National Polytechnic Theater, a building whose use was approved by the minister of the interior. Napoleón Saltos, head of the Coordinated Social Movements (CSM), was elected secretary; former Supreme Court president Carlos Solórzano from Guayaquil was among the delegates. The assembly members rejected any dialogue with the government, saying they were tired of promises and lies, and offered themselves to the nation as a democratic alternative to the existing government, whose legitimacy they had long denied. CONAIE for years had urged the replacement of all three branches of government by something akin to its own Parliament of the People.

At a January 11 news conference called to denounce the dollar plan, Vargas reiterated his plea for Ecuadoreans to reject their government as the only way to avoid economic and social disaster. He announced that CONAIE would block roads and stage other forms of civil disobedience for as long as necessary to force the president from office. In private, he and his associates admitted their chief goals were economic and social reform, with or without Mahuad. *The uprising of the Ecuadorean people is under way,* Vargas declared at the same news conference. Fellow Indian leaders, some of whom dressed in traditional multicolored ponchos and feathered hats, seconded his statements. He predicted the protest would be a step-by-step campaign in which indigenous groups would move, mostly on foot, down from remote mountain villages to provincial towns and on to Quito. Four days later, Vargas forecast that within three days the nation would be at a standstill. *We want,* he said, *to have the country paralyzed by then.*[2] Levantamiento leaders estimated that 40,000 of their followers would come to Quito but later scaled back the figure to 10,000. Many news reporters said nearly 20,000 Indians converged on the capital. The new defense minister, General Carlos Mendoza, estimated that throughout the country, 30,000 indigenous people had mobilized against the government.

The uprising promised to be a massive threat to Mahuad's administration, and on January 14 the president met with his Council on National Security in the Defense Ministry to analyze the situation. Afterward he issued his fourth state-of-national-emergency decree in seventeen months. The president assured the public that uniformed public servants would prevent the Indian movement from paralyzing the nation by blocking roads, halting traffic, and menacing public buildings. The decree authorized him to use troops against threatening crowds. He sent about 30,000 soldiers and police across the country to watch the roads and contain the Indians. The Army had emergency authorization to detain any demonstrators it wished, and hundreds of antiriot police were positioned to guard government buildings. *We have a constitution and laws, and protesters cannot act above them*, Mahuad insisted.[3] The Armed Forces rejected Minister of the Interior Vladimiro Álvarez's numerous pleas that the Armed Forces both arrest and court-martial anyone who violated the law during public demonstrations. General Mendoza deemed the request unconstitutional as well as an effort to draw the military into politics as arbitrators against CONAIE, the CSM, labor unions, and others.

The Levantamiento was for the most part peaceful and orderly. On January 15, television and radio broadcasters announced that food was scarce in the traditional Indian street markets and that traffic was greatly reduced on provincial highways. Most observers had expected to see police firing tear gas at thousands of Indian protesters and were surprised by the country's calm. While hundreds of antiriot police guarded government buildings, Quito's 1.2 million residents played soccer in the parks, strolled in the streets, and enjoyed the weekend.

Early the next week, matters worsened for Mahuad. Workers from the state oil company, Petroecuador, announced an indefinite strike on January 17. It was another labor headache on top of the trade union day of protest terminated just five days before. Upward of 4,500 oil industry employees now insisted the president step down, and their leaders condemned him for being unable to revive the battered economy and for failing to provide Petroecuador with adequate funds to compete with private foreign multinationals. Fully aware that nearly half of Ecuador's revenue came from oil exports, both Mahuad and the strikers knew the significance of paralyzing Petroecuador. The shutdown significantly lowered production. Whereas the average output in 1999 was 373,290 barrels per day, in January 2000 it fell to 260,000. The work stoppage continued until Mahuad fell, and not until January 24 did operations return to normal.

As the Parliament of the People met, Indians continued coming to Quito. Their ranks dramatically swelled on January 18 when, under cover of darkness, some 5,000 entered town. In small groups, many were disguised as Cholos to avoid detention by the military units guarding the roads and virtually surrounding the capital who, pursuant to the state-of-emergency decree, had orders to disperse any gathering they deemed a menace. CONAIE leader Vargas claimed that military personnel had forced anyone wearing a poncho to get off buses destined for Quito. *The strategy of the government was to prevent the Indians from arriving in the capital, to restrict them to the provinces.... They arrived anyway, on foot,* he said, *and it created great unity for the different actions we have under way.*[4]

Former president Rodrigo Borja condemned efforts to block indigenous people from traveling as *a completely reproachable racist measure.*[5] General Mendoza denied that many Indians were stopped by the military, since they *had arrived in the city of Quito in small groups of two, three, and up to four persons . . . a situation in which we could not intervene . . . although we could control the place where they congregated.*[6] Once in the city, the Indians assembled as planned in El Arbolito park near the capital's colonial center, close to major government buildings including Congress, the Supreme Court, and Carondelet, and across the street from the U.S. embassy. The park's ample grounds served as a base from which to stage marches.

According to Minister of the Interior Álvarez, on January 18 Mahuad asked the military high command if any of them knew Colonel Lucio Gutiérrez, who, an informer told him, planned to kidnap prominent officials and businessmen and hold them until the president resigned. Generals Mendoza and Telmo Sandoval replied that the information was erroneous, that Gutiérrez was loyal.

On January 19 policemen urged the thousands camped out in El Arbolito to return to the provinces. A hovering police helicopter dropped pamphlets that pled in both Spanish and Quichua, *Your animals need you.*[7] Unmoved by the fliers, the Indians marched on the government buildings a few blocks from the park. Riot police guarding them kept their distance as the demonstrators milled about. No one provoked violence, and no efforts were made to storm either Congress or the Supreme Court, but that same day thousands of marchers paralyzed downtown traffic, blocking streets with rocks, log, and burning tires. In the provinces, Indians disrupted major roadways and marched on the centers of towns and cities. In many communities they were joined by a broad spectrum of the population, including Cholos, students, and professionals.

Around 4 PM the next day, January 20, Mahuad was abruptly summoned from a meeting in which the dollar plan was being discussed and presented with chilling information from Álvarez. The president was briefed on the situation unfolding near Congress and the Supreme Court. Both buildings were surrounded by a circle of police, who were in turn looped by Army troops. Around the double tier of government forces were two larger circles of several thousand Indians bearing clubs and machetes. They were shouting to those who tried to enter or leave their outer perimeter, *Damn it, nobody enters, nobody leaves! (¡Carajo, nadie entra, nadie sale!)*. Salvador Quishpe, the head of ECUARUNARI, later said:

> *On the 20th we had to circle the national Congress and we spent the entire day there, encouraging the actions of our companions. Many young companions wanted to enter Congress on the 20th, breaking the circle of police. That, of course, was very difficult for us to do . . . on the one hand it was necessary to enter Congress, but on the other the moment was not right . . . but the same euphoric companions asked that as national directors we order them to enter; nonetheless, we could not do things in a precipitous manner. We had to wait and see for the most opportune moment. . . . We also had information that Lucio Gutiérrez was going to arrive with a group of his companions, and that gave us a certain hope. The problem was we were never sure whether they would arrive or not. It was the first time they were going to break the traditional scheme in which the Army had always allied with the government and the businesses that govern the country.*[8]

The president immediately summoned Mendoza, and Sandoval to explain what he perceived as a dangerous military strategy of the Indians. He was described as furious when he asked the generals why they had told him the day before that the situation with the Indians was under control, when the double ring of thousands around Congress and the Supreme Court clearly demonstrated that it was not. Disturbed by the tone and substance of Mahuad's questions, Mendoza answered that the president's attitude reflected the antagonism of the interior minister and other presidential advisers toward the military establishment. Further, he chastised Police Commander General Jorge Villarroel for not having first provided the Armed Forces, who he said were in charge of security, with the alarming information of the Indians' encircling action. Villarroel retorted that his only obligation was to keep the interior minister informed since the police fell under that ministry's control.

According to Mendoza, when Mahuad asked him what was the military's strategy to contain the Indians, he replied that its obligation was to protect both civilians and soldiers. Thus, *we based our actions in exhausting the Indian groups, avoiding any confrontations so that they would abandon their action out of fatigue and for lack of logistical support.*[9] Mendoza assured Mahuad that the situation would be contained: *If the Indians take two steps, we will take three strategic steps.* But the president demanded to know why the danger was allowed to develop in the first place, asking, *How is it possible that you cannot control a demonstration of 4,000 people?* The general answered, *Do you want us, Mr. President, to suppress these people with blood and fire, using our rifles?*

Mahuad proceeded to query Mendoza on the recent meeting between CONAIE leaders and the Joint Command: *What have we gained from your conversations with the Indians?* Mendoza answered that CONAIE wanted a new government, something that everyone had long known.[10] According to Vargas, during CONAIE's meeting with the Joint Command the generals and admirals merely listened. He later said: *we did not ask anything of the Army. What we said to everyone was that as Ecuadoreans we had to unify for a complete change of government. We did not want a corrupt parliament. The Court should dissolve itself. We said all of this with clarity.*[11] Vargas had stated as much at the news conference on January 11 and had made similar and frequent public pronouncements since March 1999.

Mahuad's adviser on Indian affairs, Diego Iturralde, joined the heated conversation and revealed that the night before (January 19), Vargas had said at a meeting in the Casa de la Cultura, adjacent to El Arbolito, that CONAIE could go ahead with its plans because it could count on military support. He never made clear what those plans were, but the implication of collusion between the Indians and the military was made when Iturralde informed the president of another matter. A few days earlier, Mahuad had granted CONAIE's request that its leaders be permitted to speak with the officials of the Joint Command. In the words of General Mendoza, *The doors of the military institution had always been open to those sectors of the population who wanted to maintain peace and the democratic system, and to end the crisis. The indigenous sector, one of the most neglected and impoverished of our country, accused civilian governments of never having resolved their ancestral problems . . . they looked for answers from the Armed Forces.*

At no time, Mendoza insisted, did the military try to *assume the role of arbiters in Ecuadorean politics and never did the possibility arise of implementing a new political system foreign to the constitution.* The president acquiesced to CONAIE's solicitation on condition that the meeting take place in the Agriculture, not the Defense Ministry, and that Interior Minister Álvarez and Labor Minister Ángel Polibio Chávez be present along with Iturralde. Contrary to Mahuad's instructions, the group of about twenty CONAIE leaders met in the Defense Ministry and without the two cabinet officials. In response to Iturralde's information, the generals told Mahuad that Vargas refused to meet with anyone if the two ministers attended. For their part, the cabinet members interpreted their exclusion as some sort of military maneuvering but were not alarmed.

During the meeting, according to General Mendoza, Vargas told the Joint Command that CONAIE *had come in the name of all the Indian community, of the poorest people of the country, and that they aspired to make structural changes in the state*. He referred to corruption *as a virus that had invaded all areas of public administration and that the only way the country could recuperate was to depose the three powers of the government that were responsible for the crisis into which the nation had fallen*. He continued by saying that *the executive power had an intimate relation with corrupt bankers*, and he castigated legislative and judicial corruption as well. Vargas stated harshly that the Supreme Court was *a place where no Indian had ever been able to assert his legal rights because all that functioned there was money*.

Mendoza then said that Vargas expressed *confidence in the Armed Forces, the only institution that had demonstrated preoccupation with the situation of the poor*. The general was told that CONAIE did not want anyone *wounded or killed* and asked only for a *just democracy*. Mendoza replied that *there existed constitutional channels for the demands of the citizenry*, and that *it was necessary to proceed in conformity with the constitution and the law*. The general later said, *At no time was there any kind of conspiracy, a plot did not exist with the colonels or with popular leaders to destroy the executive*.[12] Immediately after the hour-or-so conference concluded, thousands of Indians gathered near the Defense Ministry. Iturralde finished his remarks by saying that police intelligence had let him know that in the early morning hours of January 19 and 20, military vehicles took provisions to those gathered in El Arbolito. Armed with this new information, Mahuad exhorted Mendoza to explain what was transpiring with the Indians, to which the indignant general, in the face of what he deemed *the president's doubts*, offered his resignation as head of the Armed Forces: *I am leaving now, because I am not in agreement with what you are doing. I am tired of listening to you. I do not want to know anything because these people that you have here, this sociologist* (Iturralde), *are lying. They are not telling the truth, they are confusing things*.[13]

Mendoza returned to the meeting almost immediately, and before leaving again he urged the Armed Forces and the police to work together to avoid a takeover of Congress and the Supreme Court. To confront the crisis, he proposed creating a committee composed of the ministers of defense, interior, and finance, the president of the Supreme Court, and the commanders of the Armed Forces and police. He suggested the group convene at 9 AM on January 21 in the office of the Joint Command, with Interior Minister Álvarez presiding. Before the day was out, on behalf of the military Mendoza issued a communiqué supporting the constitutional order.

Moments before the meeting adjourned, General Sandoval promised to cooperate with police forces to send a rescue mission that evening (January 20), to let some thirty-seven employees leave the blockaded legislative and judicial buildings. Among those held were twenty-eight employees of Congress, five magistrates of the Superior Court, and four members of the Supreme Court. Galo Pico, the head of the Court, called the government several times to insist on their release. Álvarez and Pico agreed that help would be sent between 10 and 11 PM that evening. After 11 PM Pico again phoned Álvarez about aid for the trapped employees, and Álvarez in turn called Villarroel, who explained that for tactical reasons the mission had been put off until 4 PM the next day. General Sandoval, he said, made the decision as head of the Army.[14] According to Colonel Jorge Luis Brito, whom the Junta of National Salvation had named as the nation's military commander, Mendoza and Sandoval planned to dislodge the Indians and the Junta of National Salvation:

> *At 5:30 pm an urgent meeting took place in the Joint Command Center of Operations to evaluate the most recent events. General Mendoza instructed General Carlos Moncayo Gallegos, commander of Tarea Force No. 1, to act with great energy to repel the Indians . . . General Sandoval ordered that 5,000 men be moved to neutralize the action of the Indians; that they discharge a great quantity of gases to disperse them and that they use firearms in the case of extreme emergency; and that if that was not sufficient, that they fire at their legs. At the end of the meeting, General Wilson Torres Zapata, director of Army Intelligence, said he was not in agreement with the decision to repress and dislodge the Indians, and that the only solution was political negotiation.*

Brito received a phone call shortly after midnight from news reporter Bernardo Abad, asking for his opinion on removing the Indians who surrounded Congress and the Court. *I responded firmly that it would not be convenient because there were women and children, that it would produce many deaths and injuries, and that instead of solving the problem it would make it worse, with grave consequences for the stability of the country, and that it was necessary to search for alternatives.*[15]

Two decisions on the 20th by Generals Mendoza and Sandoval make it clear they had wearied of Mahuad and were willing to use CONAIE to pressure him. Permitting the Indians to form a double ring around Congress and the Court, followed by their refusal, initially at least, to risk conflict by breaking the human circles around the buildings and rescuing congressional and court employees reveal that intent. From Mahuad's vantage point, the generals clearly had not taken adequate steps to control the Levantamiento. What neither general knew was that the next day their efforts to push Mahuad from power would be enormously complicated by a group of junior officers.

Meanwhile, in El Arbolito park, shamans led a variety of ceremonies; one of the most common sought the assistance of the spiritual guide Yachag. Groups of Otavaleños and other nationalities distributed potatoes, bananas, and other food to those forming the outer circles. Careful not to overreact, Mahuad assured the growing crowds that force would not be used to remove them so long as they remained peaceful, but he urged CONAIE leaders and followers to watch out for infiltrators with dynamite and firearms. Some student groups, it was reported, had committed isolated acts of violence, and a handful of masked protesters fired guns and hurled Molotov cocktails at policemen.

Blasco Peñaherrera, director of the polling firm Market, had prophetically commented on January 15 that Mahuad had entered a crucial period. The president had to show he could revive the economy and deal with discontent, and although the dollar proposal boosted his popularity to 22 percent from a low of 9 a month earlier, Peñaherrera remained uncertain about Mahuad's future: *If there is blood, if there is violence, then I think the Armed Forces could be thinking about changing the president. . . . The crucial point is for the president to take a position of leadership. A majority of people want a strong hand.*[16] If Mahuad maintained order and got his dollar plan through Congress, the pollster predicted, he would survive into next year. But if he failed with the deputies and if the strikes and protests spun out of control, his days in office were numbered. January 21 was the critical date for Mahuad.

It was clear before January 21 that CONAIE's Parliament of the People was offering itself as an alternative to Mahuad's regime and that the Indian movement's representatives had held discussions with senior Armed Forces officials. But until that Friday the importance of the contacts initiated months earlier between CONAIE leaders and disgruntled junior officers was hidden. Antonio Vargas said that the same convention that had reelected him president of CONAIE on November 19, 1999, *made the decision to have a great mobilization for a complete change, to eliminate the three powers of government*. Vargas later summarized CONAIE's efforts to gain military support for the Levantamiento: *We prepared ourselves during November and December, speaking also, in the last days, with the colonels. First with the generals, second with the colonels, and third with the captains. Later we had three plans . . . with generals . . . with colonels . . . with captains . . . those were the alternatives. . . .*[17]

In the early morning hours of January 20, some of the several thousand Indians encircling Congress and the Supreme Court tried to enter the buildings now protected around the clock by hundreds of policemen and soldiers, but they quickly desisted when they met opposition. At about 9:45 AM, troops guarding the buildings mistakenly welcomed three buses filled with junior officers and their men from the Military Polytechnic School. However, the captains and lieutenants of the 195-man group used the buses to open a passageway in the circles of police and soldiers surrounding Congress. The gap was immediately closed by Indians who advanced toward the building and joined 150 of its guards, who were members of the Heroes of the Cenepa Engineering Brigade. These men had fought at the front during the Peruvian war only five years earlier, and they were disillusioned with Mahuad's peace concessions and their dismal pay during his budget-conscious administration. Because of their recent losses and sacrifices in defense of the nation, they commanded the respect of the police, their fellow soldiers, the Indian movement, and the general public. The Cenepa Brigade had arrived about half an hour before the Polytechnic force and joined them in taking over the building. One brigade member, Captain César Díaz, wore his silver medal awarded for fighting on the frontier and claimed to speak for his colleagues: *I didn't want to leave my children a country full of moral and economic misery.*[18] Hundreds of Indians and soldiers rushed Congress and took control of the legislature. The barricades surrounding the building were taken down and the nearby Supreme Court was occupied without opposition. By mid-morning, CONAIE leaders and junior Army officers sat in seats once occupied by deputies.

Two things were readily apparent. For one, the Indian movement had the cooperation of part of the Armed Forces. For another, the nation's military was fractured; the only question was how deeply. The day before the occupation of the two government buildings, three Army captains told Vargas the decision had been made to take Congress the next day. Shortly thereafter, the two central participants in the event spoke late at night by cell phone. During the conversation, Vargas informed Colonel Gutiérrez that the next morning CONAIE would capture the Congress building and reiterated his hope that the colonel would join in. The two men understood that the armed guards would not open fire on the Indians. Gutiérrez agreed with trepidation. He was preoccupied by what he considered too weak a military force, but he relented under pressure from Vargas, who feared losing the support of his followers if the Levantamiento dragged on much longer. Many Indians simply could not stay away from their crops and livestock for an extended time.[19] Gutiérrez was urged to act by some of the colonels and captains who had plotted with him, and he likely was aware that Mahuad had heard conflicting reports of his loyalty. Already compromised were some of the officers of the Cenepa Brigade, the Military Polytechnic School, and the Army Finishing School. Those of the War Academy were not committed, but Gutiérrez thought many would adhere. Still, the colonel feared rebel military strength might prove woefully inadequate.

Only a few weeks earlier, Gutiérrez had been assigned to the headquarters of the Joint Command because it needed an engineer with administrative abilities. He held a degree in mechanical engineering from the Military Polytechnic School and after graduation had kept in close contact with his alma mater. He had begun communications with Vargas at least three months prior to the January Levantamiento, and a large measure of trust had developed between them. The colonel was introduced to CONAIE's leaders through retired colonel Marco Miño. Until 1996, Mino had worked with the National Institute for Agricultural Development (INDA), which brought him into contact with Indian communities throughout the country. Miño's brother-in-law of two years was General Paco Moncayo, hero of the recent war with Peru and current congressman. CONAIE knew Gutíerrez's convictions, including his belief that the country was pitifully managed and his outrage against Mahuad's weak efforts to control corruption. They also knew that he was bitter about the poor treatment of soldiers in the wake of peace with Peru and blamed Mahuad for giving up to much in reaching a settlement.

Gutiérrez headed a cadre of fifteen colonels who had met intermittently for three months to discuss the possibility of a military-Indian alliance to topple the government. At first they thought the top brass wanted to remove the unpopular president, but by December 20 they were determined that if the generals failed to respond to their overtures and those of CONAIE, they would act on their own. Until January 20 they believed the generals would either lead or join a movement to overthrow Mahuad. The colonels agreed on many issues, including the need to eliminate corruption; revitalize the Armed Forces with more men, better pay, and a bigger budget; halt the plan to substitute dollars for sucres; and place limits on administrative decentralization, which they feared would divide and weaken the republic. They concurred on keeping price controls on basic items of consumption, freeing frozen bank accounts, pursuing and punishing bankers who had fled the country after pilfering public funds, and placing strict public-interest provisions in proposals to privatize public businesses. On most major issues the junior officers and CONAIE were in substantial accord. They refused to dwell on their differences over military spending and decentralization.

The officers' meetings had included one on December 20 with 62-year-old former paratrooper Lenín Torres. In 1962, along with thirty young lieutenants, Torres planned to revolt in Quito, proceed to Esmeraldas to be joined by Air Force pilots, and launch a land and air operation to take over the government. The movement was discovered and Torres ended up confined to García Moreno prison for five years. Upon his release, he traveled to Chile to support Salvador Allende's socialist regime. Torres always believed that change in Ecuador would come through an alliance between the military, Indians, rural workers, and urban labor unions. Gutiérrez and Torres had met through the latter's son Sandino, as they had served together in Carondelet during Abdalá Bucaram's presidency—Gutiérrez as an aide-de-camp and Torres as part of the palace guard. In the process of removing Bucaram from office, Gutiérrez said: *I was his Aide-de-Camp, but above all a military man with a promise to the people. When on February 5, 1997, orders were given to defend the palace and I saw there was a danger of firing against the people, I met with officers of the presidential guard and explained to them that there are occasions when the immediate situation takes precedence. We therefore decided not to open fire on the people, because all the people of Quito had assembled, representing all Ecuadoreans, and demanded President Bucaram's departure.*[20]

Along with the other conspiring colonels, the pair thought another former paratrooper, a Venezuelan who, like Lenín Torres, had spent time in prison for a failed rebellion, provided an excellent example. They admired Colonel Hugo Chávez's popular reform government in Venezuela and wanted to emulate it. The colonels met seven times after December 2 in offices, cafés, and private homes in the capital. Included in some of the gatherings were CONAIE leaders Vargas and Quishpe along with Pablo Iturralde from the CSM. Major actors among the colonels were Lucio Gutiérrez, Sandino Torres, and César Díaz. CONAIE leaders also talked to officials at the Polytechnic, including César Villacis, a retired general who taught there. Villacis often showed his students documents that he claimed demonstrated corruption in Guayaquil's customs house and frequently vented his anger over Mahuad's inaction in correcting the national humiliation and disgrace. CONAIE denied plotting with Villacis or with anyone else at the Polytechnic, and insisted that discussions with the school's officers and soldiers were limited to analyzing the country's miserable condition. As we know, crucial assistance for CONAIE's capture of the Congress building came from the Polytechnic.

According to Colonel Gutiérrez, *A group of captains, students of the Polytechnic School of the Army, and I met the night before and decided to act. The next day, at six in the morning, I gathered my family, my wife, and my daughters, and I told them of the decision I had made.*

Thus, on January 20, after Congress was occupied at about 10:30 AM, Gutiérrez left his upper-middle-class house in El Batan and drove his own car to Congress. On arrival, he proceeded to the office of the legislature's leader. It was apparent that his role had been and would be prominent. Again, according to Gutiérrez, *about twenty uniformed military men entered through one door, and through another there entered about 100 officers from the Polytechnic School.* The number, he said, soon rose to *more than 400 officers and some Army troops.*[21] Those who had adhered to the movement came from the Polytechnic School, the Cenepa Brigade, the Joint Command office, Logistical Support Brigade Number 25, the War Academy, Communications Battalion Ruminahui, and the Army Finishing School.

Their defection took most high-ranking military leaders by surprise, including the head of Logistical Support Brigade Number 25, Colonel César Vallejo. Watching events as they happened on television, he saw Colonel Gutiérrez on the screen and immediately checked his men. He found four officers and forty-two soldiers missing, among them Lieutenant Colonel Patricio Acosta and Major Jimmy Pino, the respective heads of the Communications Company and the Logistical Support Battalion. Vallejo learned that at 10:10 that morning the pair had left with a military vehicle along with a number of well-armed troops.

At the War Academy, which for the past seventy-six years had turned out the nation's top officers, those in charge were likewise taken unaware. Students and professors, after following the takeover on television, assembled to urge its director, Colonel Fausto Cobo, to join Gutiérrez and the Indians. At 11 AM, Cobo decided to take part in the insurrection, and about an hour later he was at Gutíerrez's side. A friend and protégé of Defense Minister Mendoza, the War Academy's head was respected in military circles for his intellect and leadership abilities. Within hours all the members of the Academy arrived at the Congress building but later said their decision was without advanced planning. Cobo made an important public declaration on the use of force shortly after arriving at Congress: *We are not going to fire a single shot against the Ecuadorean people, nor against our compatriots, against those officials who obey orders and come here to remove us.* The colonel received numerous phone calls from military units asking him for instructions. He claims: *I never gave a military order. When they called me I told them the following: Look, we are responsible soldiers. We are not utilizing arms. The only collaboration that I ask of you is that you not comply with orders of repression. I do not need you here, stay in your units. Had I wanted power at that moment, I would have told them to capture their commanders. It was as simple as that.*[22]

The new occupants of the buildings housing two of the three branches of government announced the end of Mahuad's administration by the will of the people. The Junta of National Salvation, composed of two representatives from the Parliament of the People and one from the Armed Forces, replaced the nation's deputies and Supreme Court justices. Antonio Vargas claimed to lead the three-man governing group. With television cameras covering events as they happened, he introduced the Junta: himself, Colonel Lucio Gutiérrez, and former Supreme Court president Carlos Solórzano. The three symbolized the union of Indians, soldiers, and the law.

Vargas later gave this summary of CONAIE's attempts to win over the Armed Forces:

> *We spoke with General Telmo Sandoval about whether they (the Generals) would form part of the triumvirate we wanted to create . . . it was three days before January 21, our plans were already public. . . . On the 20th we made contact with the War Academy and the Polytechnic School . . . also with the Heroes of Cenepa. . . . On the last day (the 20th), the generals decided against it; so too did the colonels, and those who made the decision were the captains. They told us: without colonels, without generals, we made the decision to join the mobilization Friday morning. . . . They then told Colonel Lucio Gutiérrez: "If you want to accompany us, accompany us; but the decision has already been made. At eight o'clock in the morning we will be in Congress." . . . At 9 or 10 o'clock in the morning they had more than 300 Army men with us. . . . Friday, once the Congress was taken, Colonel Lucio Gutiérrez arrived . . . then Colonel Fausto Cobo. The*

> *young colonels and the captains said they had made the decision. At first their declarations were timid, but later they made a very strong declaration. . . . Later, Colonels Lalama and Brito came, sent by the generals, to reach an agreement. They too were convinced not to retreat, and from that point on we began to organize. What we lacked was a little strategy. All was spontaneous, it came rapidly. All came in spontaneous form, because nobody believed that we could throw out the president, throw out the three powers. It was crazy, a utopia, suicide.*[23]

The huipala, with seven colors of the rainbow symbolizing the unity of the indigenous people, was raised in the legislative chamber at noon. Many of the nation's senior officers were dismayed to see the Indian flag rival Ecuador's national banner. The Indian movement's member on the Junta and the leader of both was 41-year-old Antonio Vargas Guatatuca, first chosen CONAIE president in 1996 and recently reelected to his second term. The son of Quichua and Záparo parents, he was born and raised in Puyo, Pastaza province, in the Oriente and was fluent in both Quichua and Spanish. Vargas was only eight when oil was found in 1967 in neighboring Napo province, and the racial and cultural discrimination he witnessed against indigenous people by his teachers and others from outside the area turned him into an activist for land, language, and social justice by the time he reached his mid-teens. He visited Spain to study the Basque and Catalan nationalist movements, and after his return to Ecuador he was elected president at age 31 of the Organization of Indian Pueblos of Pastaza, where he served until 1994, leading marches and protests against the expansion and environmental destruction by petroleum companies in his native province. He became Pastaza's provincial director for bilingual education in 1994, and in 1997 at age 38 was elected CONAIE's president. Vargas had been the central figure in organizing and leading the January Levantamiento, which was far from his first. He had led thousands of Indians into Quito on several occasions in 1999 when his efforts in March of that year both won rollbacks in price increases and eroded confidence in Mahuad's ability to lead. Vargas was a pacifist and embraced the philosophy of Mahatma Gandhi.

CONAIE's vice president and Vargas's close associate was Ricardo Ulcuango, born in 1966 in Cayambe, Pichincha province, in the northern highlands. A farmer by profession, from 1988–89 he taught literacy before being chosen president of the Confederation of Peasant and Indian Organizations of Cangahua in 1991. Ulcuango served as the head of Pinchincha Riccharimi; as president from 1996 to 1999 of ECUARUNARI, the most powerful of CONAIE's three components;

TWO SYMBOLS

January 21, 2000. The Indian movement's huipala is held aloft, while the national flag of Ecuador flies above the presidential palace.

and in 1996 was elected vice president of CONAIE for a term to expire in 2001. Another prominent CONAIE figure was Salvador Quishpe, born in 1971 in Zamora Chinchipe province in the southern highlands. A Saraguro of the Quichua nationality, he served in the Army, helped form the Federation of Saraguros of Zamora Chinchipe in 1992, taught school near the provincial capital of Zamora, graduated from the University of San Francisco in Quito, and attended Eastern Mennonite University in Virginia. In late 1999, Quishpe gave up studying international law at Quito's Central University to assume the presidency of ECUARUNARI from Ulcuango.[24]

Justice, or more accurately disgust with the current system that operated under that name, was symbolically represented on the Junta by Carlos Solórzano, a deputy from Guayaquil to CONAIE's Parliament of the People. The port city lawyer had served as head of the Supreme Court from August 1995 until October 1997, when he had gained national fame for prosecuting unpopular Vice President Alberto Dahik in 1995 on corruption charges. (Dahik fled to Costa Rica.) When Congress removed Abdalá Bucaram in 1997, Solórzano believed that his position entitled him to the presidency, but he failed to persuade the Armed Forces. After Alarcón replaced Bucaram, Solórzano pursued corruption charges against the new interim president, but within a few months Congress, which thought he had gone too far in his zeal, removed him from the Court. He belonged to Mahuad's DP party until mid-1997, when he resigned for ideological reasons. By 1999, Solórzano was convinced that Mahuad—like Dahik, Alarcón, and Bucaram before him—was corrupt to the core and should be arrested.

Solórzano recounted his first contacts with Vargas and CONAIE: *I met him two years ago when a group of distinguished citizens met with me and asked me to be a candidate for the presidency of the republic. I then had the opportunity to meet Vargas because it was thought that an arrangement could be made between myself and someone from CONAIE. Political circumstances did not permit my candidacy. . . . In Guayaquil they (CONAIE) asked me to head the Parliament of the People, and named me president. In that capacity I went to Quito to attend the National People's Parliament.*[25]

The Junta's military representative, whose presence challenged the authority of General Mendoza as head of the Armed Forces, was 41-year-old Colonel Lucio Gutiérrez. He received attention in March 1999 when he publicly urged his superiors to devise a plan to make the military the guardians of the nation's revenue instead of leaving the task to civilian politicians who, he was convinced, stole and squandered without restraint. While politicians plundered, the colonel claimed, the condition of the troops grew worse. They needed two uniforms and two pairs of boots each year, he said, because the annual single issue was of such poor quality as to wear out within months. Their low salaries had fallen well behind inflation. Disturbed that Mahuad had given too many concessions to Peru, Gutiérrez opposed the reductions in military spending that followed the peace accord. Like many of his colleagues, he was well aware of recent events in nearby Venezuela and said he aspired to lead a popular revolution *just like our brother Hugo Chávez*. (The Venezuelan was popular with the poor, abolished the legislature and court system which he deemed unrepresentative and corrupt, and replaced them with a national assembly and watched traditional political parties wither in response to his reforms. Chávez had argued that only by naming military officers to key government posts could public institutions be rid of corruption.) Although all of CONAIE's proposals for revamping Ecuador preceded the Chávez era in Venezuela, there were numerous striking parallels between them.

INSIDE CHANGE

Evening, January 21, 2000. Crowds wait in front of the presidential palace while inside a Junta with a colonel (Gutiérrez) is being transformed into a Junta with a general (Mendoza).

Immediately after the Junta was formed, Solórzano declared to a cheering crowd, *We have sworn before God and country that the changes you seek and that we want will come*. Mahuad, he said, would be tried on charges of corruption. Colonel Acosta, who was involved in the capture of Congress, succinctly captured the sentiments of everyone: *Enough of this corruption, it has made us all poor*.[26] Dozens of others rose in the legislature's galleries and gave similar impassioned and impromptu speeches, most of which focused on the shame and harmful economic consequences of corruption and on Mahuad's failed policies. Colonel Gutiérrez spoke to condemn unchecked corruption and the country's social and economic system as well as to predict that Ecuador would undergo either radical change or disintegration.

Vargas spoke first in Quichua and then in Spanish, and as a pacifist he emphasized the fact that a revolution had been accomplished without a single drop of blood. After he and Gutiérrez finished speaking, they raised their clasped hands in a sign of unity. But in another gesture, one that severely shook the military high command, on behalf of the Junta of National Salvation, Colonel Gutiérrez named Colonel Cobo commander in chief of the Armed Forces. General Mendoza now faced a rival claimant for leadership of the nation's military. The Junta also named Colonels Jorge Brito and Gustavo Lalama to top command posts in the Army.

Soldiers occupied congressional offices. Some contacted television and radio stations, which were asked to provide live coverage of events as they unfolded; others prepared extradition orders for bankers accused of pilfering public funds. Above all they appeared preoccupied with the issue of rampant corruption. Gutiérrez set up headquarters on the first floor of the building, and more than anyone seemed to be in charge. Congressman Paco Moncayo, the retired general who had played a crucial role in Bucaram's exit, submitted to the Junta his resignation. He said later he did not believe the junior officers had carefully planned Mahuad's ouster: *The colonels were worked up against corruption. They did not have a leader or a visible plan. It was nothing but an explosion against a terrible circumstance.*[27] Lieutenant Colonel Mario Lacano Palacios agreed with Moncayo's assessment, as did Colonel Luis Hernández Peñaherrera, who as head of the Escuela Militar Eloy Alfaro was asked but refused in midmorning on the 21st to join the insurrectionists.

For the last ten days General Mendoza had been both de facto defense minister and head of the Joint Command. On the morning of the 21st, he was informed that 8,000 Indians continued to surround Congress, and that *the possibility did not exist that they could break through the concentration and shields of armed security forces.* At about 9 AM, as members of the Crisis Committee were assembling, he was told that *the military circles had been opened and the Indians had been given freedom to enter the national Congress, an action aided by some military elements.*[28] Forty-five minutes later, Mendoza was informed that Colonel Gutiérrez had directed the military movement to aid the Indians. The general

OFFICIAL ESCORT

January 21, 2000. An elderly woman walks on the street near Carondelet, the presidential palace.

watched events on television, in the company of General Sandoval and General Villarroel. The trio responded quickly and assembled almost three dozen generals, admirals, and colonels in Joint Command headquarters. The nation's military nerve center was La Recoleta, near the presidential palace, which housed the Defense Ministry and the offices of the heads of the Army, Navy, and Air Force. After doubling security around Carondelet with 300 additional soldiers, they decided their first task was to establish a minute-by-minute monitoring of the country's response, particularly that of its military garrisons, to the takeover of Congress and the Supreme Court and the formation of the Junta of National Salvation. The first response came at 10:30 AM. The rebellion, it appeared, was confined primarily to the Polytechnic School and the War Academy. The news changed as the hours passed.

The generals decided on a twofold strategy. First, Mendoza would try to reason with the colonels occupying Congress. Second, Sandoval would communicate with Vice President Noboa, who was at his home in Guayaquil, and make contingency arrangements to bring him to the capital. In pursuit of the first objective Mendoza sent Colonels Brito and Lalama, both from the office of the Chief of Staff of the Joint Command, to persuade the rebellious junior officers to cease and desist. Mendoza, Lalama, and Cobo were all good friends, and Mendoza hoped that Lalama could dissuade Cobo from persisting in the insurrection, and that Cobo in turn could persuade Gutiérrez. Before leaving for the legislature at 11 AM, the emissaries asked Mendoza for a document authorizing them to negotiate with Gutiérrez, but the general snapped back: No negotiation; Gutiérrez must simply give up. Gutiérrez was clearly the key to the Junta, as his membership provided it with a measure of military approval and served as a magnet to disgruntled soldiers throughout the country. His continued presence could further fracture the Armed Forces.

Mendoza placed two phone calls: the first to War Academy director Colonel Cobo, just named head of the Armed Forces by the Junta; and the second to Colonel Lalama, the emissary named to persuade Gutiérrez to stop his insurrectionary actions. Mendoza asked Cobo to meet to discuss motives and how to end the crisis that dangerously divided the military. Cobo refused. He later said: *If they want to get us out of here, they will have to take us dead.* He explained the reasons why he had revolted: *Dr. Mahuad fell by himself. The people of Ecuador sentenced him. What action has he taken against corrupt bankers; has he returned one sucre to the creditors? The people have revoked his mandate.*[29] Those points aside, for Mendoza and other senior officers the young colonel's presence on the Junta as well as the conduct of others in uniform who occupied Congress threatened an intolerable fracture within the Armed Forces. Colonels Lalama and Brito informed Junta leaders that their mission was to avoid a dangerous and potentially bloody division within the military in which soldiers as well as civilians would confront each other. At noon Mendoza finally made phone contact with Lalama inside the legislature: over much shouting in the background, Lalama succinctly informed his superior: *My General, it is irreversible, it is irreversible*. Instead of returning to Mendoza's office, Lalama and Brito joined the insurrection in Congress.[30]

The nation's top generals and admirals assembled again in La Recoleta for an update on the status of provincial military garrisons. They were told that in Guayaquil, Cuenca, Loja, and Pastaza a few high officials were reported to be sympathetic to Gutiérrez and the Junta; and initial information indicated that their predilections might escalate into armed rebellion. No one knew or could know for sure. Similar reports were heard about a handful of Air Force and Navy units. Mendoza feared others would soon join the rebellion, and he was particularly concerned that the actions of the prestigious War Academy would attract military units to Gutíerrez's movement. The general was, he said, *preoccupied especially with the possibility of a division within the Armed Forces. While the Navy and Air Force still had not detected problems, there existed the danger that discontent could grow incrementally, a hypothesis that could not be discarded.*

Fearing that a major schism in the ranks might soon develop and that the unpredictable and rapidly unfolding situation could escalate out of control, the Joint Command saw no alternative but to abandon Mahuad and to communicate at once with the nation. *The military institution was put in serious danger because a person would not abandon his pride. . . . In these circumstances, the high command publicly reiterated that the Armed Forces had withdrawn its support.*

Vice President Noboa had already been contacted by the high command, and transportation and security were provided to bring him to the capital. At 12:30 PM the top brass met with the press where General Mendoza, in writing, called upon the president to provide the country with what he called a *constitutional response* to the crisis. Clearly the Armed Forces wanted Mahuad to resign. Mendoza and the generals then met with Interior Minister Álvarez, who queried Mendoza on a proper *constitutional response*. Mendoza replied he would tell the president in person. Álvarez also inquired about the fate of the rebellious colonels, and Army commander Sandoval assured him they would all be subject to military law.

Mendoza and top military men met with Mahuad to explain why he had to resign. The general offered this account of the confrontation: *When the situation became complicated, we told the president he had to make a decision. You make a decision, we told him, to avoid trouble in the country. Try to pacify things. He banged the table. He lost his composure. And I told him, you cannot pound on the table. You are facing the military command. You are wrong*. Mahuad refused to quit. *I am not going to resign!, and if this is a coup d'état I will not resign!* The president, according to Mendoza, then *ordered that they proceed to dislodge all of those found in Congress. Several of his ministers supported him and urged the use of force.*[31]

Seeing no other solution, Mendoza and his entourage marched out of the room and at about 3:40 PM reassembled at La Recoleta. General Carlos Moncayo, brother of Congressman Paco Moncayo, in an even more explicit statement on behalf of the Joint Command urged the president to relinquish his office as the only means of ending the crisis.

Mahuad became desperate. He tried to contact Vice President Noboa in Guayaquil to urge him to condemn the coup and support the administration. The idea came from Álvarez, who placed the call himself only to be told that Noboa was meeting with his own advisers, was following the situation closely, and would speak out at an opportune time. General Sandoval had already contacted the vice president with the instruction from the military high command that he prepare to come to Quito on short notice. Though he showed no signs of betrayal to Mahuad, Noboa was well aware of the dangerous implications of a division within the Armed Forces. At 4 PM former president Osvaldo Hurtado again urged Mahuad to resign, and around 5 PM, although refusing to resign, Mahuad left his office in Carondelet. The Joint Command had withdrawn his military escort as a message that he would not be protected in the presidential palace. (Abdalá Bucaram had received the same signal on February 7, 1997, when the military withdrew most of his personal protectors.) General Carlos Moncayo, moreover, had already told Mahuad he should abandon the building.

The president's strategy for quelling the insurrection was largely devised by his adviser, Carlos Larreátegui: to condemn the coup and hope that U.S. government pressure to maintain democracy would be sufficient to keep him in office. In a television broadcast to the nation, Mahuad declined all demands to step down and declared that anyone who wanted to overthrow him must use force. In a televised interview that followed the address, Álvarez added: *The president has no intention of leaving the country.*[32] Mendoza and the generals had earlier tried to talk him into exile and put a plane at his disposal at Quito's Air Force base to take him anywhere he wanted to go, but Mahuad refused to board or to yield his office. Shortly thereafter, an army unit surrounded the plane, apparently under orders to detain Mahuad and his ministers. Mahuad left the air base, took refuge in the Chilean embassy, and from there went to the home of a friend before leaving the country. He had not resigned. *A thrown-out president does not resign,* he told the nation. *He is thrown out.*[33]

As soon as the president left his office at about 5 PM, General Mendoza entered it to take over the executive's duties. He subsequently said he did so because the vice president was in Guayaquil. *I felt that someone had to fill the vacuum,* he said, adding that *the Armed Forces on January 21, 2000, had to elaborate a gradual strategy, in accord with the evolution of events.* The general sent a telegram to all military units announcing that the Armed Forces had taken control of the nation. Events moved rapidly early that evening. Colonels Brito and Lalama reiterated to Gutiérrez that the Armed Forces were threatened with a grave division, and they urged him to leave the Junta to restore institutional unity. A basic demand of the Levantamiento and the Junta had now been met with Mahuad's ouster. At that point, Gutiérrez considered the appeal but hesitated to make a decision. Mendoza remained preoccupied. *Some Army units had demonstrated signs of sympathy for the movement of the colonels. Others maintained an air of expectation,* he later said, although *the majority accepted the orders of the high command.*

Learning that Mahuad had "abandoned" Carondelet in the wake of the Joint Command's insistence that he resign, thousands of Indians showed signs of moving from Congress to the true center of power. Controlling the legislature and high court, they lacked only the executive office to dominate all three branches of government, each one of which they proposed to abolish. Around 6:30 PM, Vargas, Gutiérrez, and Solórzano left for Carondelet, surrounded by a crowd of Indians and soldiers. Gutiérrez was already thinking of leaving the Junta to avoid bloodshed; Vargas and Solórzano wanted nothing to do with the Joint Command that claimed to be in control. Gutiérrez continued to consult with Colonel Cobo and finally decided to talk to Mendoza and other senior officers in Carondelet. Mendoza was anxious to discuss matters with the colonels, and he later said the high command had to act with *sufficient tolerance to not produce greater divisions . . . between members of the Armed Forces.* CONAIE and the Junta, Mendoza believed, meanwhile *incessantly sought the adhesion of other military units to consolidate their position.*[34]

Civilian opposition to the Junta of National Salvation formed rapidly on Friday, January 21. Hostility came fast and furiously from the U.S. government and international organizations such as the United Nations and the Organization of American States (OAS). Countries in the region and several in Europe weighed in rapidly against what they deemed a destruction of democracy. From Friday afternoon onward, General Mendoza felt mounting pressure from both home and abroad to get rid of the Junta, and by the end of the afternoon he believed that it could not survive.

THE JUNTA WITH A COLONEL

January 21, 2000. From left to right, Carlos Solórzano, Colonel Lucio Gutiérrez, and Antonio Vargas of CONAIE. The young colonel's presence begged the question of the role of senior officers in the Armed Forces and in the nation.

Significant civilian hostility to the takeover of Congress by soldiers and Indians began almost immediately, in part because television cameras relayed the events as they happened. By early Friday afternoon a group of politicians, businessmen, and others banded together to insist on a constitutional approach to replacing Mahuad, whom they all wanted to resign. As with Bucaram's removal in February 1997, the issue was not whether the president should leave but rather who would replace him: the Junta of Vargas, Mendoza, and Solórzano; Vice President Noboa; or someone else. Former president Hurtado assumed a leading role and condemned the Junta as an unacceptable dictatorship. If the crisis were not constitutionally resolved within three days, Hurtado promised that on Monday, January 24, at 10 AM in the Avenida Los Shiris, all-out demonstrations would restore the constitutional process. CONAIE's own tactics would be used against it.

There are brigades and elements in distinct sectors of the Armed Forces, Hurtado said, *who are not backing this, and accordingly nobody can believe that it is consolidated. Who is going to back such a ridiculous, such an absurd thing? Today we have the proclamation of a dictatorship, not by the head of the Joint Command of the Armed Forces but by Dr. Solórzano. Whom does he represent? Can we permit Ecuador to be governed by a citizen who represents only himself and his ambition and interests? We are not going to permit that.* Hurtado later would label the events of January 21 *a ridiculous, shameful epilogue.*[35]

In Guayaquil, Mayor León Febres Cordero watched the takeover of Congress on television. As his first action, the former president called the commander of the First Naval Zone in the port city, Vice Admiral Fernando Donoso, and the commander of the Military Zone of Guayaquil, General Oswaldo Jarrín. He told both men to maintain order and to inform their troops there would be *an enormous spilling of blood if they broke with the democratic process.* The mayor asked the vice admiral and general to declare Guayaquil's independence from the illegally formed Junta in Quito. He then called a press conference where he told reporters he would ask the president of Congress to convene an extraordinary session the next day to oppose the coup and restore democracy. He wanted the deputies to meet in the *loyal* city of Guayaquil.[36]

Meanwhile, foreign pressure against the Junta mounted. Political stability and Armed Forces unity in Ecuador were important to the U.S. government for more than ideological reasons, as upheaval could only weaken the country's response to the guerrillas and drug cartels threatening to spill across the border from Colombia. Economic interests were also at stake; the United States was committed to free markets and neoliberal policies throughout the hemisphere, as stable economies, liberal trade, and investment would benefit its own economy. General Mendoza spoke to U.S. diplomats, including Ambassador Gwen C. Clare, at least three times on Friday afternoon. The general described the conversations as calm and brief with none lasting longer than a few minutes, adding that he let the diplomat know *we were going to protect constitutional democracy in Ecuador, but we weren't sure how.*[37] He revealed that during the short talks these officials threatened Ecuador with isolation and assured him the U.S. government would cut off foreign aid and discourage investment unless constitutional procedures were followed and the Junta disbanded.

The warnings were public as well as private. Peter Romero, the State Department's top official for Latin America since September 1998, could not have been more blunt when he made a radio broadcast aimed at the Junta from Washington on January 21. Ecuador, he announced, *had to continue on the path of constitutional democracy or risk facing isolation, not only from the United States and from Europe but from the whole international community*. If Junta leaders persisted with their unconstitutional regime, they could expect *political and economic isolation carrying with it even worse misery for the Ecuadorian people*, who would *face political and economic isolation akin to Cuba's.*[38] Romero had visited Ecuador about five weeks earlier, had been the U.S. ambassador to Quito from 1993 to July 1996, and was well acquainted with the country's leaders. Despite this pressure, Mendoza insisted that *the United States did not influence in any way the decisions of the high command.*[39] CONAIE head Vargas disagreed: *The people wanted a change, but larger countries always manage coups, and the United States gave the order that this process would not continue*. Colonel Gutiérrez concurred with Vargas, stating that Mendoza felt *pressure from the North American embassy and from Guayaquil. . . . The embassy wanted to implement dollarization and privatization in a savage manner. . . . The same with the coastal oligarchy. . . . These pressures were too strong for General Carlos Mendoza.*[40]

State Department officials rapidly mobilized regional pressure against the Junta.[41] An emergency meeting of the OAS was called in Washington on Friday afternoon, and delegates adopted a resolution that condemned the coup and expressed full support for Mahuad. Weeks later, the secretary general of the OAS, César Gaviria, would explain that the international community wanted to make it clear that *the level of isolation and economic harm to the country would be so great as to make the Junta unsustainable.*[42] (Gaviria earlier had served as president of Colombia and was a champion of democracy, human rights, and judicial reform. As president he formed a peace agreement with four major rebel groups.) In another regional action, South America's Mercosur trade bloc comprising Brazil, Argentina, Paraguay, and Uruguay as well as associate members Chile and Bolivia, issued a joint statement that implored Ecuadoreans to respect their democratic institutions. Chile not only condemned the coup but also offered asylum in its Quito embassy to Mahuad and to *any Ecuadorean democrat* who needed it.

The United Nations' Secretary General Kofi Annan admitted on Friday afternoon that he was following events in Ecuador closely, and he was *firmly convinced that the best interests of Ecuador and its people can only be served through the maintenance of constitutional order and the rule of law*. Swift condemnation also came from Europe. British officials denounced what they called an unconstitutional coup and called for Mahuad's restoration. Foreign Office minister John Battle elaborated: *Since the era of military-led regimes in the 1970s and 1980s, Latin America had been moving in the right direction toward entrenching democratic values. The news of an unconstitutional coup in Ecuador involving the military is a sad development and a setback. It is vital to the economic interests and stability of the whole region that democratic government be restored in Ecuador as soon as possible*. On the same day a French Foreign Ministry spokeswoman said in Paris: *France deplores the interruption in the normal functioning of institutions and hopes to see a rapid return to constitutional order. France calls on all parties to respect democratic institutions and thus demonstrate their spirit of responsibility*.[43]

For years the Armed Forces had championed constitutional government and resisted the temptation to take power. As political leaders became discredited by continuing corruption and as social and economic conditions worsened, a majority of Ecuadoreans looked to the military as the institution they could trust and support. An August 1999 Cedatos poll showed 65 percent of the people with a favorable impression of the men in uniform, whereas politicians tallied a mere 6 percent. During the last several decades, the Armed Forces had gained considerable credit for improving the lives of the poor, having provided healthcare and educational facilities to many long-ignored and impoverished communities. The military had been one of the first in Latin America to assume a developmental role, largely for the purpose of thwarting guerrilla movements. Its role in community development was clearly acknowledged in the Law of National Security and the Organic Law of the Armed Forces.

Not only was the Army involved in a host of development activities, but also the Navy put up schools and parks in poor districts of Guayaquil, rebuilt many communities, and constructed numerous houses on the coast after El Niño's course of destruction. Many observed that the social origins of the soldiers stood in contrast to that of most other armed institutions in Latin America. The officers were not upper class but were mainly middle-class urban men, while most of the rank and file came from the rural lower middle class. They had a reputation for honesty, which set them apart from the public's perception of most politicians. Former U.S. ambassador Leslie Alexander, who left his post in July 1999, offered an index to corruption. He noted that most politicians wore gold Rolex watches, while the majority of the generals told time with Seikos. His views mirrored those of the public at large.[44]

In early 2000, a major problem for Mahuad was the particularly low regard in which he was held by most men in uniform. A foreign diplomat in Quito who spent considerable time analyzing the institution reported: *To a man, they have no respect for him. And it amazes them to have to stand up to their fellow citizens and defend someone whom they view as corrupt and inefficient.*[45] Many soldiers were Indians from impoverished backgrounds, and some battalions were more than 50 percent indigenous. Colonel Cobo said of the Indians and the Army: *The Indian people had their units, consisting only of Indians. But one should not forget either that our conscripts, that our soldiers are really from the community and that many of them, the majority, are Indians. The average soldier is a poor person . . . because of that the Ecuadorean people have great affection for us.*[46]

Ecuador had 35,000 professional soldiers in 2000, and enlisted men earned the equivalent of $40 per month while colonels garnered about $250. General Mendoza himself received $640. One colonel involved in the takeover of Congress claimed his pay had dropped from the equivalent of $1,000 to $250 per month. After January 21, stories spread in some circles that many of those involved in the takeover of Congress saw taking power as a means of climbing out of deep personal indebtedness. Members of the Armed Forces suffered a huge decline in income during Mahuad's administration, and most resented the fact that military resources and personnel shrank after peace was negotiated with Peru. Mahuad recognized that *one of the consequences was that the military budgets were cut*. In fact, they were declining in most of Latin America, including Peru. While both Mahuad and his Peruvian counterpart praised the peace agreement for freeing funds for social programs, General Mendoza lamented that reductions in the budget had forced the military to mothball some of its hardware and to put on hold its plans to modernize seriously outdated equipment. Most of those in uniform thought the reductions had gone too far, despite the fact that similar reductions were taking place throughout the region. Young officers railed against the apparently unchecked corruption that permeated politics, and they blamed stealing and government mismanagement for the reduction in their salaries by inflation and devaluation.

The indigenous population thought they endured the harshest effects of their country's economic troubles. Mahuad and most politicians, they believed, lacked both understanding and concern about the Indians' situation.[47] Soldiers who participated in building roads and hospitals in remote provinces doubtless knew more about the plight of rural Indians than the nation's political elite. Quito analyst Felipe Burbano observed what he called *a growing distrust among sectors of the military toward the way civilians manage things. Perhaps the Armed Forces,* he surmised, *experience the country's sufferings more.*[48]

Not everyone was happy with the developmental role of the military. In June 2000, Febres Cordero, reflecting on the military during the 1970s and the events of January 21–22, 2000, urged a reform of the curriculum for training the Armed Forces:

> *I do not believe that human contact between the Armed Forces and the social realities of the country does the institution damage, but a soldier is a human being like any other. The problem lies in the study curriculum of the Armed Forces, where instead of studying war, as one should in the war academies, they study political development that does not involve the Armed Forces. They have now returned, in one way or another, to what happened in the decade of the 1970s when instead of dedicating themselves to the study of problems vital to territorial security and the state, they dedicated themselves to becoming planners. As a consequence we had the dictatorship of General Rodríguez Lara and the problems we have had with the Armed Forces in this decade.*[49]

At some point before January 21, Indian leaders and junior officers came together. General Mendoza acknowledged that days before the coup attempt he was well aware that Colonel Gutiérrez was conspiring against the president. General Gallardo, whom Mendoza replaced as defense minister on January 11, remonstrated that Mendoza knew weeks and not days in advance that junior officers planned to support the Indian protest, and they berated him for acting irresponsibly in not moving decisively to discourage insubordination.

The head of the Army, however, had acted. When General Sandoval received a report from Army intelligence on January 17 about discontent in the ranks, he made a surprise visit to the Military Polytechnic School. He listened to disgruntled officers' complaints about the country's worsening economic condition, their low salaries, and unchecked corruption and to their insistence that the military remedy these and other critical problems. Sandoval told the malcontents that the Joint Command would respect the constitution. These comments reportedly left his listeners disillusioned and more determined. A few days after Congress had been captured, Gallardo claimed those and other junior officers were motivated by an idealistic desire to save Ecuador from economic and social chaos. *Young men are passionate,* the general explained. *When they see injustice, they want to take action imme-*

diately to end it. They don't have the experience and vision of older people.[50] January 21 showed that the military was a force divided between the younger, more radical junior officers and the older, more conservative top brass. The immediate challenge for the high command was to prevent the fracture from worsening and to prevent bloodshed.

THE JUNTA WITH A GENERAL

Evening, January 21, 2000. From left to right, Carlos Solórzano, General Carlos Mendoza, and Antonio Vargas of CONAIE. The general's presence partially answered the question of the role of senior officers in the Armed Forces and in the nation.

Of immediate concern on January 21 to General Mendoza and the Joint Command was word that the Junta and its Indian supporters, who controlled Congress and the Supreme Court, were readying to march on and seize the presidential palace. Vargas explained: *The people applied pressure to take the presidency, saying the presidency is where one governs . . . and at four in the afternoon, ex-President Mahuad had left the palace.* Senior military officials ordered well-armed soldiers behind the sandbags that surrounded the building. Thousands of Indians entered the area around the palace, but instead of trying to seize it, they celebrated. Soon the Indians and others amiably mingled with armed soldiers in combat gear. CONAIE's president later related his surprise at the assemblage of the Armed Forces elite inside Carandolet: *It was not in our program that they would be there, and only when we were advancing were we told that the generals were in the presidential palace. When we went to the presidential palace, at seven in the evening, we learned that the generals were inside. All the generals were there from the four branches: Army, Navy, Air, and Police. . . . We entered and we said: We have come only to take control, not to negotiate.*[51]

Mendoza and other top military leaders conferred for several hours with the three members of the Junta of National Salvation. Vargas told Mendoza: *The chief of the Joint Command is Colonel Fausto Cobo; Generals, your reign has ended. The government has already been formed.* Mendoza and the other generals, Vargas reported, *told us they were in agreement with the process of the government of National Salvation, of the Junta, but that they were not in agreement with Colonel Lucio Gutiérrez being part of it.* Mendoza presented Gutiérrez with two pages entitled, *Official Proposal of the Armed Forces and the National Police.* The document declared that the Armed Forces and Police should take power to restore order and peace and pledged a prompt restoration of democratic institutions. Vargas and his colleague Quishpe responded with dismay at the document they thought negated a role for the Indian movement and kept the country's power structure in place. They said they wanted nothing to do with any of it and suggested the generals and admirals leave. Mendoza remained and reiterated the point that Colonels Brito and Lalama had made in their earlier mission to Colonel Gutiérrez: it was essential to avoid civil war by maintaining military unity and not dividing the country. Therefore, the Junta's junior officers must recognize the high command. On the crucial question of where the locus of power would reside, the general and the rest of the high command decided it rested with them.

Discussions followed. Solórzano, Vargas, and other Indian leaders met in Mahuad's office while the generals and admirals gathered in a nearby conference room. Mendoza ordered Brito and Lalama to again ask Gutiérrez to leave the Junta, and used his friend Colonel Cobo to help facilitate an agreement. Mendoza later said that *Colonel Cobo, surely because of the grave situation in which the War Academy found itself, and reflecting on military hierarchy and discipline that could not be separated, came up with a new possibility for restructuring the Junta of National Salvation. He proposed including General Carlos Mendoza as a member of this provisional government.*

Mendoza thought the suggestion *had the possibility of a way out of the conflict.* Admiral Enrique Monteverde oppossed the move, saying: *I am not in agreement with any of this, and right now I want to be relieved of duty.* The admiral changed his mind, however, when Mendoza privately asked him for his support and told him: *This is temporary.* He explained his strategy: *To accept and to step down.* The heads of the Army, Navy, and Air Force—General Sandoval, Admiral Monteverde, and General Hernán Batallas, respectively—all agreed to the chief of the Joint Command of the Armed Forces, General Mendoza, replacing Gutiérrez as a member of the Junta.[52]

Gutiérrez finally relented and agreed to be replaced by Mendoza, saying: *Nobody has to convince me of anything. I am certain that the Armed Forces must maintain unity.* With these words, he obeyed orders. According to Gutiérrez, *As I had made clear as a condition for my participation in the movement, all had to be completely nonviolent, so I believed it was a good alternative . . . but I obtained the promise of the colonels and also of General Mendoza that he was not going to defraud us, that he was going to move ahead with the process that Ecuador wanted.* Gutiérrez later recalled that *the supposed insubordination, if it was that, was an error immediately corrected because later we subordinated ourselves to the high command and gave it control of the situation . . . the high military command removed its backing and asked for the resignation of the president, later removed his security guards, and today we have another president.*[53]

The generals and colonels emphasized to a disappointed Vargas that it would be impossible to govern without military cohesion, and that civil war and bloodshed must at all costs be avoided. Vargas told Mendoza: *if it were guaranteed that the process would go forward, then there would not be a problem.* Vargas accepted the general as a Junta member. Gutíerrez's decision to step down left Vargas with little choice but to relent to Mendoza, or so it seemed, as he was now without even a facade of military might. The CONAIE leader later recalled: *The colonels and the captains said, "Antonio, there is going to be bloodshed here." The colonels that were with us were unarmed, their hands were totally empty; on the other hand, the generals were totally armed. Nevertheless, the people were down below. . . . Colonels Cobo, Brito, Lalama, Lucio, and the captains said that . . . Mendoza was an honest man, and that, because of that, they had decided.*[54] In any event, Vargas never intended to resort to violence or risk bloodshed, although, as he later confessed, he was not adverse to a show of force as a bargaining chip to advance the interests of the Indian movement.

The accord was reached minutes before midnight, and the four past and present Junta members joined with those who controlled the nation's armed might to participate in an old ritual. In addition to Vargas, Solórzano, Gutiérrez, and Mendoza, those in uniform who joined hands in prayer were General Sandoval of the Army, General Ricardo Irigoyen of the Air Force, Admiral Monteverde of the Navy, and General Villarroel of the Police. According to Quishpe, *We sang the national anthem, we prayed the Paternoster, but in the end, it was all a hoax.*[55]

Mendoza, Vargas, and Solórzano emerged on the balcony of the presidential palace and presented themselves to the throngs as a new triumvirate. Vargas and Solórzano remained as members of the reconstituted governing body. The presence of Mendoza alongside the other two Junta members was meant to send a strong message to the country, namely, that the Armed Forces stood united behind the movement that toppled Mahuad. Some in the crowd shouted for Gutiérrez to speak. He did not. Solórzano delivered an animated and emotional speech in which he referred to the stepped-down colonel as a valiant man who would likely be the next interior minister. He told the thousands gathered there that the Junta would make political and economic changes to benefit the masses of people, lashed out against corruption, and pledged not to retaliate against the officers who had participated in the insurrection.

Vargas then spoke to the multitude in both Quichua and Spanish to assure them that the Armed Forces and the police had guaranteed to continue the process begun that morning. He closed by declaring: *Ecuador triumphed, you triumphed, we triumphed. Long live Ecuador!*[56] After the reassuring speeches, Independence Plaza in front of Carondelet virtually emptied. Fewer than twenty people remained to mill about, and the majority of Indians retired to El Arbolito. Yet most of Vargas's associates were disappointed in the switch of Mendoza for Gutiérrez, and one wryly commented that upon hearing the news, he immediately had *the same sensation as in February 1997.*[57] The reference was to the minimal changes that had occurred three years earlier, following massive protests and demonstrations, when Alarcón replaced Bucaram as the nation's head. Bucaram himself claimed that the so-called oligarchy had manipulated the Indian movement, Pachakutik, and other progressive forces to topple him, only to take over the government and rule largely through the conservative PSC.

Junta members proceeded to speak to the press about the present situation and their future plans. Mendoza claimed the full support of the military and could therefore guarantee freedom for all citizens. When members of the triumvirate were asked what they would do about Mahuad's plan to revive the economy by scrapping the sucre for the U.S. dollar, they replied that they would discuss the move. The response was surprising, given the vehement opposition to the plan on the part of Vargas and Solórzano. In answer to most questions, however, Mendoza deftly repeated the same words: *Tomorrow, tomorrow we will respond to all of these concerns.*[58]

Meanwhile, anxious members of the Joint Command gathered in La Recoleta to await an explanation of what had happened. Some discussed Solórzano's proposal to name Colonel Gutiérrez interior minister, a suggestion that ran counter to the senior officers' sense of hierarchy; it was more than customary for top-ranking officers to receive the post if indeed it went to a military official. The Police were also agitated, as tapping the colonel meant they would be under the command of an Army official. Everyone thought it urgent that Mendoza explain what was going on. At least twenty regional commanders, moreover, made it abundantly clear that they adamantly opposed a military takeover. They wanted to follow the constitution with civilians in charge.

Mendoza, Vargas, and Solórzano gathered in the early morning hours of January 22 in an annex to what had been Mahuad's office. A little after midnight, Vargas began the meeting by saying: *Starting now we begin to govern. The first thing we have to do is decree the confirmation of the Junta and the lifting of the state of emergency*. While Mendoza sat silent, Solórzano expounded on what decrees should be issued. He insisted the dollar plan would be discarded, bank accounts would be unfrozen, and the exchange rate would be fixed at 14,700 instead of 25,000 sucres to the dollar as Mahuad had proposed. Vargas later said: *We worked on decrees for more than 45 minutes, and the only decree made was to not recognize the three powers of the state and to close all the airports for awhile so that the thieves could not leave.*[59]

Within minutes, Mendoza rose and announced: *Gentlemen, excuse me, I will return soon*. A quarter of an hour later he was in the Defense Ministry to bluntly give the Joint Command the explanations they so eagerly awaited. *Gentlemen, here is my resignation. At this moment I retire; all is prepared for dissolving the triumvirate*. Before departing, Mendoza made General Sandoval chief of the Joint Command. *I leave you*, he told Sandoval, *in charge of the Armed Forces*.

Mendoza immediately returned to Carondelet and at about 2 AM on January 22 informed Vargas and Solórzano of his exit from the Junta. *Gentlemen, this body is dissolved; I renounce my position, I am leaving. My family is not in agreement*. Vargas and Solórzano were incredulous; both begged Mendoza to remain for a month, a week, or a few days at least. Solórzano asked, *And now, do I have to go too?* to which Mendoza replied, *What else remains? The two of you are alone*. The remaining Junta members were left without the vital backing of any armed element. Vargas and Solórzano were told, Vargas related, *that the situation had changed, that they had decided to support the vice president, that the vice president would assume the presidency because that was, they said, democracy*. Vargas's response was, *You were traitors to the people. The people had great hopes for change*. Mendoza went home at about 2:30 AM and gave the press an account of what had transpired. He had, he said, completed his objective *to maintain the unity of the military institution and to guarantee the presidential succession.*[60] The Indians who had been assembled in front of the presidential palace had already dispersed. By 2:30 AM there were only a few street children and beggars in the Plaza Grande in front of the building; they were accustomed to sleeping there. There were no guards, no soldiers. All was tranquil. Only a small group of Indians was near the Congress building.

LEANING AWAY

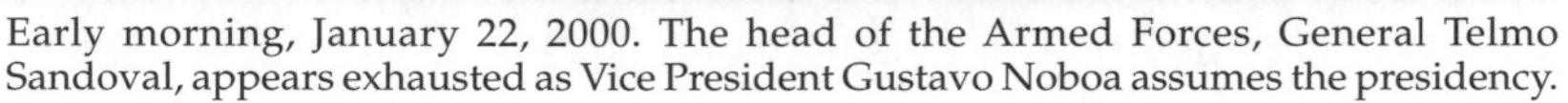

Early morning, January 22, 2000. The head of the Armed Forces, General Telmo Sandoval, appears exhausted as Vice President Gustavo Noboa assumes the presidency.

In the meantime the nation's new military head, General Sandoval, convoked a council of thirty-six generals and admirals and told them the military schism had been healed: the constitution would be respected, and Vice President Noboa would assume the presidency. Arrangements had already been made to bring him from Guayaquil to the capital. Shortly after his arrival on January 22, Noboa took office in the Defense Ministry at 7 AM. Vargas later emphasized that *the current president took over in the Joint Command office, not in Congress, not in the presidential palace*. Jorge Loor, leader of the National Peasant Council, said that within a matter of hours Ecuador had made a radical change in governments. One *had the legitimacy and representation that is the expression of Ecuador's great majority, which was manifest in the Junta of National Salvation made up of Lucio Gutiérrez, Antonio Vargas, and Carlos Solórzano*. The other *declares itself legal and is led by Gustavo Noboa, supported by the right-wing parties, bankers, and big business.*[61]

Mendoza subsequently gave several reasons for temporarily replacing Gutiérrez on the Junta and joining with Vargas and Solórzano. According to the general, he allied with the insurrectionists when it became clear that they might storm the presidential palace. Had he not done so, the guards at Carondelet would have responded with gunfire. As head of the Armed Forces his duty was to avert bloodshed as well as to avoid military fractures, no matter how devious the means. Faced with the uprising of Indians and colonels, he felt obligated to restore order and maintain as much military cohesion as possible. Those goals required him to feign loyalty to deceive the Junta and to quickly sacrifice himself and resign from the Armed Forces. Staying on, Mendoza explained, would have made him a target for the anger of dissident officers as well as the Indians, which would have been contrary to the urgent need to restore institutional and national unity. To do other than manipulate and abnegate, he projected, *would have brought on a social explosion*. From the outset, Mendoza assured everyone, he planned to dissolve the Junta and hand over power to the vice president, thereby respecting the constitution and restoring military unity. Thus, *I decided to negotiate a peaceful solution to the situation by agreeing to become a member of the Junta. I then asked my companions in the Armed Forces to reestablish the juridical order through a presidential succession.*[62]

Mendoza's motives for joining the Junta immediately became a matter of controversy. He had his version, while others offered differing accounts. Some believed he hoped to retain power for himself, was thwarted by the Joint Command and provincial commanders, and accordingly was suddenly converted to following the constitution and letting the vice president take charge. The general's own account was accurate but incomplete. What Mendoza omitted discussing were two decisions on the 20th made by him and General Sandoval that made it clear the pair had wearied of Mahuad, wanted him out, and were willing to use CONAIE to pressure the president into departing. Their resolve to let CONAIE form a double ring around Congress and conceal it from the president until it was an accomplished feat, followed by their refusal to risk conflict with the Indians and break the human circles around the building to rescue congressional and court employees, suggest to some that they had that intent. Colonel Brito asked, *Was perhaps taking Congress the condition necessary to force the resignation of Mahuad? Surely that scenario had been foreseen in the meetings at the end of the year in Balneario de Punta Blanca.*[63] Brito was referring to meetings between Vice President Noboa and Generals Mendoza and Sandoval at Noboa's home in Guayaquil. The determination of the generals on the 20th to let CONAIE enhance its strategic position around Congress gave yet another signal to disgruntled junior officers that the top brass wanted the president to step down. What neither Mendoza nor Sandoval knew, however, was that the next day their efforts to push the president from power would be greatly complicated by junior officers. The actions of Gutiérrez and others of lesser rank in aiding the Indians presented the senior officials with a dangerous, delicate, and unexpected situation, namely, an intolerable schism within the Armed Forces.

What had happened to Ecuador's democratic institutions and processes? When Congress convened after the Junta was disbanded, it condemned what it called a coup d'état. OAS leader Gaviria painted a broader picture, labeling it *a rebellion of a large number of citizens aided by a conspiracy with elements of the Armed Forces at officer level.*[64] General Mendoza's summed up his role: *For me the greatest treason would have been to divide the Armed Forces*. When he announced the end of the short-lived Junta, he said he had briefly joined the body only to *prevent a bloodbath.*[65]

Many analysts saw it otherwise. Some contended Mendoza gave power to the vice president only because he did not have sufficient military support to keep it himself. Retired general and recently resigned Defense Minister Gallardo questioned Mendoza's depiction of himself as a defender of democracy, and he suggested that he and Sandoval, as respective heads of the Joint Command and the Army, took advantage of the growing discontent of junior officers to further their own political ambitions. They stood in stark contrast, Gallardo believed, to the young officers who were motivated by idealism. The pair knew of the conspiracies weeks in advance and failed to take steps to check them; instead, they sought to use the young men to further their own ambitions. Other analysts shared Gallardo's appraisal. One political commentator said that only when Mendoza and Sandoval failed to get the support of the Joint Command and of at least twenty regional commanders, *did they suddenly become democrats.*[66]

In an interview, Mahuad concurred with Gallardo's assessment: *I was overthrown by a military movement that started plotting weeks earlier. It is what they speak of as a military coup in Ecuador. Many participated in it, among them those who today speak with their mouths full of democracy.*[67] Mendoza countered Mahuad and other critics. *If I had had ambition, I would have stayed that night and taken advantage of the situation.*[68] He said he sought a bloodless as well as a constitutional solution to the crisis that would forestall further military disunity.

The remaining members of the Junta, Vargas and Solórzano, were overwhelmed and, Vargas claimed, deceived by Mendoza and the top brass. Both denounced dissolving the Junta and permitting the vice president to assume power, and each continued to urge Mahuad's arrest. The fallen president remained free, and although Colonel Gutiérrez was promptly detained by military personnel, Mendoza appealed for compassionate treatment for the colonel and for the other officers implicated in the coup d'état. Angered that the military manipulated his movement, CONAIE's President Vargas said, *We never had any desire to take power. We knew that if we tried to do so, there would be a lot of bloodshed.*[69] He claimed Mendoza had promised to share power with the indigenous movement, only to betray it. Neither the military nor the Indians got what they wanted. Once the country's most respected institution, the Armed Forces now stood fractured and reeling. The indigenous groups had forced out Mahuad but in his place got his vice president, who was likely to continue conservative public policy.

Vargas later elaborated on his perspective of what had happened, citing what he considered his movement's mistakes as well as its successes:

> *There was great pressure from people who wanted to take over the presidency because it was another symbol of corruption. Perhaps it was a mistake, and if we had stayed in Congress the results would have been different. Maybe it was a mistake to accept Mendoza, but if we hadn't, a great deal of blood would have been spilled, and neither the colonels nor the indigenous leaders were willing to take that risk. . . . If we had been stubborn about power, there would have been confrontations and death. But since we weren't, we offered the possibility of another path.* On the positive side, Vargas concluded: *The important thing is that we caused a shakeup of the political class, and it was made clear that there is a rebellious population here that will make change without violence. . . . We achieved our objective of bringing down the three government branches, if just for a few hours. But above all we were able to put the branches' lack of prestige in the center of debate. Many said this was a fantasy, suicide, a crazy idea of the leaders, but we proved them wrong. . . . From this experience, a new military is born, which sees that change is possible. It is the hope that the generals do not have, because they are more involved in corruption and defending their interests. There is a seed that one day will germinate throughout the Armed Forces.* Reiterating CONAIE's pacifism, its leader maintained: *Change will arrive without violence, peacefully, directed at the communities. What occurred on January 21 was a rehearsal. It made people feel in their hearts that it is possible, that they must not remain quiet because it only serves to help the same politicians as always.*

Vargas later concluded that he and his associates made three major mistakes on January 21.

> *First, we should have governed only from Congress. We did not have to leave for anything, only to take the means of communications, ministries, order the closing of airports, name a cabinet, and we didn't do that. The second mistake . . . we did not have to negotiate; we failed when we went to negotiate. The third mistake was that the people should have guarded the presidential palace, the colonels too; they should have guarded it until all was consolidated. Because in some provinces, for example, in the city of Guayaquil, the people still were not fortified; on the other hand, in other provinces nearly all the people had already risen; they seized the government; the Army and the Police were already with the people.*[70]

Were Mahuad and Gallardo right? Did the high command callously use the Indians and junior officers in a cynical campaign to depose Mahuad? Did they intend to grab power themselves? Were they thwarted only when faced by overwhelming threats from regional military commanders, the U.S. government, and the international community? Or were they taken by surprise by the junior officers and put in the position of having to respond rapidly to avoid bloodshed and further fracturing of their institution? It seems unlikely that a conspiracy of senior officers directed the events of January 21–22, that they would have risked social and institutional upheaval to topple Mahuad. Domestic and foreign opposition would have been too much to withstand, and that fact was fully understood by midafternoon Friday. It is equally improbable that the high command would have attempted such a move without first gaining the approval of regional commanders, at least twenty of whom opposed disrupting the constitution.

The timing of events likewise argues in favor of Mendoza's general version, as does the decision of Mendoza and Sandoval made well before noon on the 21st to fly the vice president from Guayaquil to the capital. Speculation about Mendoza's ambitions likely stemmed from the brief interval between his replacing Gutiérrez on the Junta and explaining his actions. Anxious regional military commanders were doubtless caught confused by the rapidity of events. Mendoza and Sandoval, who had hoped to use the Indian movement to remove Mahuad, were surprised when CONAIE and the junior officers allied to use each other for the same purpose and to create an unanticipated split in the Armed Forces. Just as Mendoza and Sandoval did not anticipate the actions of the junior officers, neither did CONAIE expect the top brass to ease out them and the colonels once they were rid of Mahuad.

On February 16, OAS secretary general Gaviria looked back on the events and concluded that *President Mahuad did a great service to democracy when he recognized that his functions had been terminated, and asked for support for the government of President Noboa . . . he facilitated the international community's interpreting of what had happened as a transition to a new constitutional government.*[71] Judged by the standard of recognizing and adjusting to harsh realities, CONAIE's President Vargas, who had acted to avoid bloodshed and assented to General Mendoza's replacing Colonel Gutiérrez on the short-lived Junta of National Salvation, likely did a great service to his country as well.

8

Gustavo Noboa

Eighteen hours after the coalition of dissident junior officers and indigenous protesters formed the Junta of National Salvation, it became clear that 62-year-old Gustavo Noboa Bejarano would be Ecuador's sixth president in four years. A century and a half earlier, his great-great-grandfather Diego Noboa had served as president for eight months. More recently, in 1996, his brother Ricardo had run an unsuccessful campaign for the nation's highest office.

On January 22, 2000, Congress met in the Central Bank in Guayaquil and passed a motion stating that Mahuad had *abandoned his position*.[1] Vice President Noboa would serve the remainder of Mahuad's term, which would expire on August 10, 2003, and four days later the head of Congress formally invested Noboa with the presidential sash. The deputies also elected a new vice president, Deputy Pedro Pinto Rubianes, a 69-year-old engineer from Mahuad's DP and the only member of that party to call for the president's resignation months before he fell. Most considered Noboa and Pinto conservatives, and the country's two center-right parties formed a majority in the parliament. DP and the PSC together commanded sixty-three of the 123 congressmen if their party blocs remained united. More than two-thirds of the deputies confirmed Noboa and Pinto.

Prior to his election as Mahuad's running mate, Noboa served briefly in 1983 as the appointed governor of Guayas province and managed a sugar refinery for a short time, but his career was more academic than political, as he taught for more than two decades in Catholic high schools before becoming rector of the University of Guayaquil, where he also served as head of the Law School. Noboa's reputation was that of a gifted negotiator and a warm, calm, easy-to-talk-to man who spoke clearly and frankly. As vice president he had overseen the reconstruction of highways and bridges on the Pacific coast that El Niño-generated storms destroyed in 1997–98. The new president described himself as a political independent, but most observers considered him to the right of center.

The new executive tried to project the image of an honest leader determined to tackle major problems and to modernize the nation. He enjoyed early support. In order to act before it dissipated, Noboa called on legislators *to recognize Ecuador's harsh reality* and to make the difficult decisions needed to put the country on the right track, insisting: *I'm not going to waste time; Ecuador is running out of time.* Noboa stressed that he was not a professional politician and thus was untainted by the scandals and political maneuvering that had discredited his predecessors; he claimed freedom from obligations to any particular constituency or special-interest group. He denied all ambition and with assumed modesty announced: *I've never aspired to the presidency, but now that I am here, I will do what I have to do.*

Noboa's principal adviser, apart from his brother Ricardo, was 47-year-old Juan José Vivas, with the post of General Secretary to the President, who later said that as Mahuad made error after error, Noboa had stayed loyal, even though Vivas warned him to distance himself from Mahuad's mistakes. The initial reaction of most politicians was supportive.

Noboa pled for patience and understanding from foreigners and Ecuadoreans alike and assured them that he was serious about restoring *confidence and credibility*. He described the tumultuous events that preceded his swearing-in as *buffoonery* and with a measure of chagrin stated that *Ecuador is not a banana republic. We are going to modernize and change this country,* he promised, *both politically and economically.*[2] To that end, on February 2 he named his brother as head of the National Council of Modernization (CONAM). Analysts soon credited Ricardo with being the administration's chief policymaker, whose ideas framed approaches to modernization, privatization, and decentralization. To head the Ministry of Energy and Mines, the President appointed mining businessman Pablo Terán, who announced he would seek more private investment in the mining, oil, and electricity industries and would push for constructing a second pipeline to double the nation's capacity to export its oil. On February 1 the national Chamber of Commerce and Industries had issued its list of demands, namely, approval of the dollar plan, policies to stimulate production and restore business confidence, debt restructuring, greater flexibility in dealing with workers (particularly in distinguishing between part time and full time), approval of salary increases to raise consumer demand, and fixing the deeply indebted social security system. The president agreed with the businessmen and pledged to address those issues and more.

At the onset of his term, President Mahuad's reputation was that of an efficient administrator free of corruption, but by the end it was that of just another compromised politician. Noboa, by contrast, knew he was deemed honest, and he moved immediately to champion the anticorruption cause. He had no hidden wealth or offshore bank accounts, he swore, and he pledged that a top priority was to bring dishonest officials to justice. Unlike Mahuad, he vowed to extradite accused bankers from abroad, and in April he would ask the United States to revoke the visas of those who had mismanaged or stolen state funds. *Pirating,* he explained, *is exactly what a group of directors of the Banco del Progreso did when they filled their bags with millions of dollars.*[3] Francisco Huerta, the interior minister, recognized that essential to Noboa's success was his credibility. If he lost, it he would doubtless fail: *The country will no longer support even one drop more of corruption. We have to guarantee ethical conduct.*

Mahaud's freezing of bank accounts came at a high political cost, and Noboa's release of the funds on February 7 won him popular praise. *We took this decision in order to avoid having the nation's economic crisis deepen and to help reactivate the economy,* his finance minister announced.[4] About 90 percent of the $3.2 to $3.8 billion was unfrozen in February, with the balance to be paid back in installments of dollar-denominated, three-to-five-year bonds. Unfreezing the accounts was one thing, but paying account holders proved to be quite another. Those in charge of doing it, the Deposit Guarantee Agency (AGD), simply had insufficient funds.

Another major decision was to embrace Mahuad's plan to adopt the U.S. dollar, a measure Noboa thought essential to restore economic order and initiate a recovery. In mid-February the proposal was sent to Congress as an emergency measure; the legislature had only thirty days to act. If it failed to approve or reject the bill, it would automatically became law. At the same time, the new president proposed salary increases to restore lost purchasing power.

The U.S. government's position was reassuring to the new administration, and Noboa immediately made it clear how he wanted to work together with Washington. An editorial in *El Comercio* noted his eagerness to obtain U.S. loans backed by oil guarantees. His approach was supported by Congress head Juan José Pons of DP and by Foreign Relations Minister Heinz Moeller of the PSC. All knew that the loans were inextricably tied to fundamental financial reforms.

On February 1, Noboa named his own people to the top security slots. Admiral Hugo Unda was put in charge of the Defense Ministry, General Norton Narváez headed the Army, and General Mario Cevallos replaced General Jorge Villarroel as commander of the 22,000-member Police. Three Police officers were implicated in the events of January 21 and were ousted from the force: Lieutenant Colonel Fausto Terán, Major Gonzalo Suasnavas, and Ensign Luis Erazo. Noboa knew his government faced a daunting challenge: *We have to have order and discipline. After twenty years of democracy, we can't have people doing whatever they please.*[5] Several hundred officers who helped capture Congress on January 21 were detained in Quito, and commanders in the provinces launched efforts to learn the degree of their subordinates' involvement. Over 100 soldiers were ultimately detained for insubordination, among them Colonels Lucio Gutiérrez, Fausto Cobo, Gustavo Lalama, and Jorge Brito. Gutiérrez was taken into custody on January 24, the others three days later. Still more were transferred to distant parts of the country, including 150 captains, lieutenants, and ensigns assigned to various units in the Amazon who left Quito in mid-February.

Those arrested faced up to eight years in prison if found guilty of charges that ranged from insubordination to insurrection. The new head of the Joint Command, General Telmo Sandoval, pledged they would all receive fair trials. General Carlos Mendoza, who resigned from the Armed Forces on January 22, agreed that the accused should be tried but urged that they be treated with sympathy. Jacinto Velásquez, who had unsuccessfully sought the presidency himself in 1996, was appointed defense attorney for Colonel Gutiérrez and the other officers accused of misconduct. He argued that his clients had not engaged in sedition and were guilty at most of a lack of discipline. Since none of them had considered using firearms, he reasoned, they consequently had not participated in a coup d'état. If any legal remedy for their actions was in order, he suggested it be Mahuad's restoration, even though over 90 percent of the population was elated that he had been thrown out of office.[6]

A military court cautiously deliberated whether or not General Mendoza himself would stand trial. Noboa sent a strong signal on the subject on February 25 when he awarded a medal to the general. The military court subsequently refused to hear charges against Mendoza, ruling that at the time of the coup he was minister of defense and that therefore only the Supreme Court had jurisdiction to hear the civil case. That civilian court maneuvered around the problem as adeptly as had the military tribunal. Its president, Galo Pico, refused to take the case on the grounds that Mendoza was not legally the minister of defense on January 21, even though he claimed to have been, since according to the constitution members of the Armed Forces on active duty could not hold the post. The legal initiative was returned to the military tribunal, but they refused to accept it.

Article 108 of the constitution allowed Congress to grant amnesty by a two-thirds vote for *politically motivated crimes* when it was *justified by a transcendental motive,* and article 109 halted penal proceedings against those who were pardoned. All Pachakutik deputies urged a pardon for everyone accused of sedition. The PSC and DP opposed the motion and emphasized the need to restore military discipline. For its part the PRE asked that in exchange for its twenty-two votes the amnesty be broadened to include its leader, former President Bucaram, who continued to live in Panama because of a warrant for his arrest. Without a majority in Congress, CONAIE and other amnesty supporters asked that the question be put before the nation in a popular referendum.

Meanwhile, the fate of the civilians who took part in the events of January 21 remained unclear. On March 29 the Supreme Court charged Junta of National Salvation members Antonio Vargas and Carlos Solórzano with *actions against the security of the state.*[7] If found guilty, Vargas, Solórzano and the others faced three to sixteen years in prison. Former ID deputies Paco Moncayo and René Yandún, whom fellow lawmakers expelled for submitting their resignations to the Junta on January 21, were also indicted. That action, the congressional majority reasoned, meant that the former deputies took part in the uprising.

On April 12, Noboa asked Congress to grant amnesty to the soldiers and civilians involved in the bloodless overthrow of Mahuad. In a letter to the deputies, he wrote: *Amnesty for these soldiers and civilians will help peace and tranquility in our country, and contribute to creating a new spiritual climate, a vision of the future. It was not an armed coup and it was not bloody: its main protagonists were hundreds of unarmed Indians, some civilians, and a group of Army officers of varying ranks who did not use weapons.*[8] Former president Febres Cordero countered that his PSC opposed pardons because *those who subverted the order of the state—who could have generated a civil war with unpredictable consequences for the country*—had to be punished for their dangerous and treasonous acts, or the country was inviting more of the same.[9]

Ecuador's military was clearly a divided institution before and even after January 21, although how much so was open to question. Civilian adviser to the Armed Forces Luis Proaño evaluated the situation with a caution: *We can't say there won't be another coup if we continue with the very grave economic and social difficulties.*[10] And *El Comercio* editorialized on January 23: *The strong movement of sergeants and other officers is still there.*[11]

The events of January 21 demonstrated at least three fundamental things about those in uniform. First, most top military officials did not want to return to the tradition of taking power and holding it for months or even years. Second, despite their determination to unshackle themselves from a history of intermittent intervention, soldiers still strongly influenced politics. In 1997 with Bucaram's ouster, and in 1999 with Mahuad's removal, the Armed Forces were clearly the final arbiters of political life. General Mendoza made plain what most people continued to understand: that the military would act decisively to avoid bloodbaths and would do what it deemed necessary to prevent institutional divisions. And third, like the broader society, the Armed Forces were fragmented, with a considerable number of current and former members putting issues of reform ahead of institutional unity.

Indigenous leaders expressed disappointment with the breakup of the Junta of National Salvation and Noboa's ascension to the presidency, but nonetheless over 5,000 Indians started leaving the capital on January 22, vowing to return or resume their protests in the provinces if called upon by CONAIE's leaders. Quito quickly returned to its relatively calm pace, but CONAIE'S's initial words to Noboa were harsh and threatening. Moreover, ECUARUNARI's President Salvador Quishpe expressed pessimism about the future: *We don't accept the presidential succession. Mr. Noboa wants to take advantage of our people's fight to keep helping the same people as always, the corrupt bankers. We will defend our historic fight.*[12]

CONAIE's Antonio Vargas, amid rumors of his impending detention, declared he would not flee from an arrest order. He asked only that authorities not treat him as a common criminal: *I am here, showing my face. If for fighting against corruption, against misery and hunger, they choose to take me to prison, I will go.* Drawing a distinction between himself and the image of traditional politicians, he added: *I am not a criminal nor am I corrupt.* Vargas severely chastised General Mendoza, saying he had promised to share authority with the indigenous people only to betray them. *In my culture,* Vargas protested, *we keep our word.* He was convinced that Mendoza had wanted to seize power for himself and had failed owing to the dispute between himself and those who insisted on adhering to the constitution. He said of Noboa: *Thanks to the indigenous people he is President. The indigenous people want to have a dialogue, a direct dialogue, about profound changes. We want to discuss privatization, the foreign debt, dollarization. We are going to fight peacefully.*

On February 1, when CONAIE called an official halt to the January 15 Levantamiento Indígena, it proceeded through its provincial parliaments with the immediate goal of obtaining sufficient signatures to call for a national referendum on May 21. If Noboa did not work effectively to end poverty and corruption, there would likely be *a great social explosion that could lead to civil war in three to six months. And that would not be our fault. . . . We are going to continue the fighting. We may return to march on Quito and we could be a lot more hard-line when we mobilize again.*[13]

Noboa's view of Mahuad's failures in dealing with the indigenous movement was that *the second Indian uprising should not have had any reason for being. . . . Mahuad did not comply with any of the offers he made during the first protest, nor with offers he made afterward.*[14] Noboa implied that the Indians could trust him to keep his word despite obvious policy differences. Just as the business community promptly presented the president with their demands, so too did CONAIE, in a seven-part proposal to be voted on in a national referendum. CONAIE had long wanted to restructure the state, and Vargas emphasized that *here in Ecuador corruption has installed itself in every branch of power, and the Ecuadorean people know it. Congress is a great center of corruption. In making laws, it does everything for a fee.* The Indian organization gathered signatures to meet referendum requirements to put the following questions to the citizenry, asking whether: (1) bank accounts should be unfrozen; (2) Congress should be closed and a process initiated to form a new governmental body that would ensure greater participation by all elements of society; (3) the Supreme Court should be politically independent, as opposed to having justices elected and removed by Congress; (4) a twenty-year moratorium should be placed on payment of the foreign debt; (5) dollars should replace sucres; (6) limits should be put on privatizing the economy; and (7) a political amnesty should be issued for the junior officers and others who participated in the January 21 capture of Congress. *If they (government officials) love democracy so much,* Vargas stated in calling for a referendum, *they should return sovereignty to the people in a plebiscite and prove they are truly democrats.*[15]

CONAIE's suggestions provoked reactions both from within and without the administration. Noboa met the first demand in February—unfreezing bank accounts—and in mid-April he announced he was in favor of amnesty. But he disagreed with the Indians on the remaining issues. On March 1, Ecuador formally converted its currency from sucres to dollars. Like many others, including most of the country's newspapers, Noboa was certain the proposed plebiscite was unconstitutional and that the nation's charter would have to be revised to allow for the kind of government restructuring proposed. CONAIE disagreed and argued that a national vote on anything was permitted if sufficient signatures were obtained. Most people in the new administration weighed in against CONAIE's proposal; and most political parties, with the exception of the PRE, announced on February 8 they would oppose it. Their central argument was that referenda had to be called by the president with prior congressional approval for the purpose of constitutional reform, not policy decisions.

Reacting to the country's political and economic crisis, Thomas R. Pickering, U.S. assistant secretary of state for Latin American Political Affairs, arrived in Quito on February 15, at the same time as César Gaviria, secretary general of the Organization of American States (OAS). Pickering met privately with Noboa and afterward pledged that the U.S. government would help secure loans from the IMF and other lenders as soon as essential economic reforms were enacted: Gaviria talked with Noboa as well as with congressional leader Pons, former presidents Hurtado and Durán Ballén, CONAIE leader Vargas, and directors of the Armed Forces. The OAS head explained: *My presence is a sign that the people of the Americas feel what is in danger is not only democracy in Ecuador, but democracy in all of the Americas. . . . The proper functioning of democracy is the first step in order for Ecuador to overcome its problems and institutional crisis, but the country has to enter a process of rectification. . . . The country has to modernize because it is being left with the worst of everything.*

Gaviria offered help with the anticorruption program already developed by the OAS and promised to facilitate dialogue among Ecuador's conflicting interest groups. Specifically addressing the Indian movement, he urged moderation: *It should be clear that with a lagging rather than a growing economy it will be difficult for Ecuador, for example, to attend to the well-justified demands and aspirations of the Indians. They should not radicalize their position, because that would represent an enormous risk for democracy.*[16]

Both the U.S. and OAS representatives made proposals that fell into six categories: (1) fundamental financial reforms to obtain foreign loans; (2) OAS help to stem rampant corruption; (3) keeping the military in its barracks and out of politics; (4) establishing ongoing dialogue among major interest groups; (5) moderation of the Indian movement lest the economy and political system be further weakened; and (6) addressing the needs of the indigenous community.

Noboa did not need outside advice to address indigenous needs. For weeks he had conducted informal talks with Indian leaders, using Francisco Huerta and Marcelo Santos as intermediaries. Foreign Minister Moeller signaled a conciliatory approach in early February, saying: *In the short term, if the people who have suffered too much and been too patient do not see some results regarding better quality of life, then everything is going to explode.*[17] After $2 billion in new loans from international lenders was arranged in March, Noboa and CONAIE reached tentative accords on health care, education, housing, and land rights. The president proposed the creation of an Indian Fund with money from the Inter-American Development Bank and $10 million from the government. CONAIE had long asked for the agency. In return, the administration urged the Indians to keep the peace; they were not to block roads, paralyze public services, or engage in another Levantamiento. Unlike his predecessor, Noboa was at least open to dialogue.

Despite the talks and pressures to moderate positions, on at least two central issues—the dollar plan and CONAIE's proposed plebiscite—Noboa remained separated from the Indian movement as well as from most other citizens. A mid-February Cedatos poll revealed that only 27 percent of Ecuadoreans favored substituting dollars for sucres. The tally showed a significant change from January 10, when 59 percent approved of the proposal; now most people feared the switch would make them poorer. Fully aware of the numbers, legislative insiders nonetheless claimed that Congress would pass the plan. Although the same poll found that 74 percent of the public backed CONAIE's referendum proposal, the president and an overwhelming majority in Congress opposed it. *One of the problems we have in the country is a lack of representation,* observed Fernando Carrión, director of the College of Latin American Social Science, *which produces a rupture between people and politics.* Ecuadoreans, he added, affiliate more easily with popular organizations than with political parties.[18]

On March 1 the deputies of the center-right parties voted to substitute dollars for sucres, and a few days later Noboa signed the measure into law. U.S. currency replaced the sucre for all but small transactions, and over the next six months Ecuadoreans would exchange $1 for every 25,000 sucres presented to the Central Bank; interest rates of 9.35 percent would be established on deposits, and 16.82 percent on loans. The country cashed in most of its foreign reserves to finance the switch, leaving only about $150 million as a hedge against another economic crisis. Stanley Fischer of the IMF said that Ecuador undervalued the sucre and pinned the dollar very high against it when dollarization was implemented, which meant that domestic prices would go up. At the same time, the National Institute of Statistics and Census set the poverty line at $112 per month. In December 1999 the average worker's income was $53 per month. Noboa raised the salaries of the Armed Forces, Police, teachers, and other public sector workers in May to $120 per month.

Most observers agreed that for dollarization to work, it was critical for international lending agencies to approve new loans. At the same time, Congress passed the Economic Transformation Law (Ley Trole I) with the support of a conservative coalition of almost three-fifths of the legislators. Those against the bill argued strenuously that the nation had lost its ability to regulate its own monetary policy and would have little fiscal recourse if another crisis hit. Proponents countered that the country already had proven incapable of handling monetary policy in the face of too many crises.

Pursuant to Ley Trole I, the budget deficit could not exceed 4 percent of the gross domestic product in 2000 and was capped at 2.5 percent thereafter. Controls on foreign currency transfers out of the country were lifted. A minimum-wage provision of 50 cents per hour was included; the legal monthly minimum had been $53 per month. The legislation gave management more flexibility with labor, and unions argued that it weakened them and made it easier for employers to fire workers and hire less expensive temporary, part-time employees. The law opened oil, electricity, and telecommunications to extensive foreign investment. Up to 51 percent of the state's interest in the last two sectors could be sold. Privatization would proceed in telecommunications, the hydrocarbons sector, power generation, shipping, airlines, and water purification plants—in all, one-half of the state's holdings. The president was authorized to issue a concession for building a new oil pipeline, and the superintendent of banks obtained added authority to control and audit the nation's banks, which would be properly capitalized.

With these provisions in place, the World Bank forecast 1 percent economic growth in 2000. The economy had declined 7.5 percent the year before. Passage of the package brought the government and the IMF closer together, and by mid-March it was clear that new loans would be forthcoming. The IMF's Fischer commented: *They did get a pretty serious law on dollarization and restructuring the economy.*[19]

It all happened with amazing speed following the events of January 21. With Mahuad toppled, the military and civilian politicians jolted, and U.S. government promises of a bailout if serious financial reforms were adopted, within a few weeks Congress ended its gridlock and passed an economic package that met the requirements of international lenders. Its defenders and detractors alike called Ley Trole I the most important piece of legislation in fifty years. The international lending community found the new economic strategy credible, and on March 9 a group of creditors led by the IMF put together a $2-billion loan package from the Andean Development Corporation, an Andean-based development bank; the Inter-American Development Bank; the World Bank; and the IMF.[20] The money was to be made available over three years.

Foreign Minister Moeller explained the administration's view of both the benefits and deficiencies of dollarization. It would lead to stability and modernization, and economic progress would come through privatizing state-run industries and further opening markets. He acknowledged that the dollar-for-sucre swap would likely lead to an increase in prices and that salaries would not keep pace. Privatization would likely mean rising unemployment for a time in an already desperate job market. *Look, this is something that in the short term is going to have social and political costs*, Moeller explained, *but we had to do it for our children, our children's children. . . . We are putting ourselves in a straitjacket to impose discipline. We have been undisciplined. And that will change*. Gladys Silva, 29, a camera shop owner in Baños, Tungurahua province, had a different perspective: *The presidents come and go around here. It never really changes anyway—all the leaders are the same, that is to say, corrupt and bad. I've come to think that we are living in a country of the damned.*[21]

The Indian movement and the CSM vowed to thwart the dollar plan and push for a referendum, yet the March 21 demonstrations in Quito to oppose dollarization drew only about 2,000 workers and Indians, and the protests hardly lived up to early promises that they would bring Quito to a halt.[22] Another opponent of adopting the dollar, former president Hurtado, took a different stance once the plan was formally adopted: *I was not in favor of dollarization, but it is under way, and one must do what one can to advance it; because if it fails, what happened in Ecuador last year will be only a pale reflection of what will happen the moment it collapses.*[23]

Reflecting on his first 100 days in office, Noboa said he had *to make various decisions, but if the country does not understand, I will have to send them the videocassette of January 21, so they can say if that is the country they want*. About half the people agreed, as the president maintained a popularity rating hovering between 45 and 50 percent throughout his first 100 days. Dollarization, the IMF agreement, Ley

Trole I, plans to privatize and cut subsidies, and amnesty for those involved in toppling Mahuad marked Noboa's first months on the job. He wanted to halt inflation, reduce unemployment, stop the flight from the country of people seeking jobs, construct a second oil pipeline, and reform education. As for the Indians, he hoped that *the Indigenas find a nonpolitical conduit for their aspirations and that they realize them.*[24]

Meanwhile, oil prices continued to climb. By early March 2000 they had increased almost threefold in fourteen months, reaching a nine-year high just as work was finished to expand the capacity of the Trans-Ecuadorean pipeline from 330,000 to 390,000 barrels per day. Negotiations, moreover, were under way with private bidders to build an additional pipeline that would at least double the country's capacity to export oil. During the first trimester of 2000, oil revenues grew 214 percent as the price per barrel rose to an average of $24, whereas a year earlier it had averaged $8.50. The volume of exports was also up about 14 percent. Tax income from January through April shot up 149 percent as oil prices soared, and the country's trade surplus advanced by the same percentage.[25]

During Noboa's first three months in office, as we know, Ecuador replaced its sucres with dollars and won a commitment from international lenders for $2 billion in loans if austerity measures were adopted. The timing and severity of subsidy cuts and price hikes, however, was left for the future, as the president sought stability by putting them off, unfreezing bank accounts, and dealing with amnesty for those involved in Mahuad's fall.

The second 100 days saw the president change course to pursue moderate, as opposed to severe, austerity measures. He continued to face strikes and protests. Major military shifts were made because of the delicate issue of amnesty and the vexing questions of who did what during Mahuad's removal from office. There were clear signs of moderation in the Indian movement as well, at least temporarily. Meanwhile, local elections were won primarily by the left in the highlands and the right on the coast, and the amnesty issue was finally settled. New problems flared up as neighboring Colombia's drug-financed guerrilla movement showed increasing signs of spilling over into Ecuador.

In late April the government and CONAIE failed to formalize a tentative agreement to improve healthcare, education, housing, and land rights. CONAIE claimed the government reneged on its prior promises and at the last minute changed financial clauses in the texts on education and housing. The government, it appeared, wanted greater flexibility in its financial commitment. Vargas broke off talks with the new administration, and Interior Minister Huerta resigned as a result. Vargas continued to speak of *the possibility of an indigenous uprising,* explaining: *Now the government is making war against the people. The dialogue is broken because the indigenous people have not been taken seriously. . . . We have risen in great mobilizations in order to be listened to. That is how in 1990 we achieved a partial solution to the land conflict; in 1994 over the agrarian reform; and in 1998 for inclusion of our collective rights into the constitution.* In reference to Noboa, Vargas said: *We believe the government will fall and any government against the Ecuadorean people will fall.*[26]

CONAIE's Parliament of the People met in Quito in early May. It had collected at least one million signatures to present to the electoral tribunal for a plebiscite pursuant to article 105 of the constitution. CONAIE and the CSM also said they would ask the Supreme Court to try former President Mahuad for having unconstitutionally, they claimed, denied citizens access to their own money.

Supreme Court head Galo Pico agreed to the last request on May 10. If Congress concurred, he said, then Mahuad and Ana Lucía Armijos, his former finance minister, would face trial for having frozen bank accounts in 1999. According to the Court, *The crime involved is that of having violated freedoms and rights guaranteed by the constitution, particularly those of property and work.*[27] The action pleased CONAIE. But about a month later, Congress rejected CONAIE's plebiscite signatures on the grounds that the constitution prohibited referenda on restructuring the state, which it claimed the Indian organization was proposing by asking for the elimination of the government's legislative and judicial branches.

In late April there were indications that moderates were gaining strength in the Indian movement. ECUARUNARI held its 15th congress with several hundred delegates from thirteen provincial federations. Luis Macas and Antonio Vargas, the past and current presidents of CONAIE, attended, as did national deputies from Pachakutik. Salvador Quishpe, ECUARUNARI's president, seen as one of the most radical of the indigenous leaders, stood for reelection. During the ECUARUNARI Congress, Quishpe emphasized: *I have to underscore that the objective is not to overthrow a president of the republic, but to go beyond that. The objective is how to change the political and economic structures of the country so they are free of corruption and to help rebuild the country. And if in this process of seeking a profound change it is necessary to change a representative of the old political class, we will do it, but that will be only part of the process.*[28] In response, *El Comercio* asked in an editorial how the change would come about: *Through ballots? A plebiscite? With another attempt at a coup?*[29]

ECUARUNARI's delegates voted against a Levantamiento for the time being, and Quishpe lost his reelection bid for the presidency to 38-year-old Estuardo Remache of Chaupipomalo, Chimborazo province, by a vote of 122 to 100, with a third candidate receiving forty-eight votes. With the change in leadership, ECUARUNARI ceased to be CONAIE's most radical arm, and it was thought that it would seek consensus and peace during Remache's three-year term. Its new head characterized his approach by saying, *It is not useful to make demagogue-like statements. I think that ECUARUNARI is better served by being more coherent.* The focus, he said, would shift to the fundamental needs of the movement's members, and the organization would make every effort not to make precipitous decisions on political matters. Remache claimed that the issue of amnesty, for example, was of secondary importance, well behind the social and economic problems confronting indigenous people, with whom he pledged to keep in close contact through thirty-five radio stations throughout the region.[30]

Other ECUARUNARI directors were more direct in their evaluation of outgoing Quishpe, calling him impulsive in his decisions and statements to the press and characterizing Remache as a calm, unpretentious man without political ambitions. Marco Murillo, president of the evangelical FEINE, added his moderate voice in early May: *We do not want to overthrow governments but instead change economic and social policies. We have the vision of coming to power in the medium or long term, but through democracy rather than force.*[31] And in an interview on May 12, President Noboa gave his analysis of the conflict between radicals and moderates within the Indian movement:

> *I believe that a group of directors have changed the discourse since the government of Borja to the present. . . . I see a radicalization of certain groups, a dispute over who can be the most radical, because there is a struggle for leadership. . . . I have met on three occasions with groups that form the very base of the organization. They are interested in their needs, not in the issues of their directors, whose issues are that Congress, the courts, and Noboa leave their houses. . . . Mr. Quishpe has called me a traitor to the country because I signed an agreement over the* [U.S. air] *base at Manta. What ignorance! The agreement was signed by Mahuad and not by me. They asked me why I did not immediately correct him, and I said that if I had, it would have made him a hero, and that as a result he would have won the ECUARUNARI election.*

Noboa continued: *Do they want power? If so, they should take to elections as should every Ecuadorean, where there are no preferences or privileges for any Ecuadorean.* When asked about the charge that the electoral system does not permit equality of opportunity for Indians, he answered: *That is not true. There are several Indian deputies. They have been able to become deputies, and others are mayors. I believe they will win many local elections.* [32] And in late May, Vice President Pinto joined in distinguishing between radicals and moderates, calling national Indian leaders *obstinate and foolish* in contrast to the more moderate ones on the local level. [33]

Moderation was the hallmark of the first Indigenous Chamber of Commerce, which began functioning in mid-April. President Noboa was present at its inaugural ceremony. Under the leadership of Mariano Curicama Guamán of Guamote, Chimborazo province, the Chamber's objective was to eliminate middlemen and thereby obtain better prices for Indian products. Curicama Guaman emphasized productivity: *I say to my fellow countrymen that they should work with a business mentality. Only when we have demonstrated our capacity to work will we have the capacity to reclaim our rights.* On the issue of adopting dollars and abandoning sucres, he sounded a different note from most indigenous leaders: *They have slapped me in the ears for saying this, but I believe that it is one of the few ways out that remain for our country.* [34]

While the Indian movement moved toward moderation, it continued its opposition to the government's austerity proposals. To make dollarization work and to bring expenditures into greater balance with income, Noboa vowed to curb subsidies that made gasoline, natural gas, electricity, water, and basic foodstuffs artificially cheap. In early April the cost of popular bus fares were doubled, to about the equivalent of eight U.S. cents. The move was met with ineffectual demonstrations, largely from students. In Quito, after some 200 protesters threw rocks at riot police, at least thirty students were arrested after being dispersed with tear gas. Some 41,000 public employees went on strike from April 16–26, seeking and getting cost-of-living increases to keep up with inflation. Indian federations blocked the highway in Otavalo in late April to protest the proposed elimination of subsidies as well as the already enacted higher bus fares. Students, labor unions, the CSM, and CONAIE all condemned Noboa's austerity program, calling his agreement with the IMF economic slavery for Ecuador, and they vowed to fight it with marches, demonstrations, strikes, and blockades.

On Labor Day, May 1, some 10,000 Indians marched through Quito's streets protesting the IMF's austerity program and the decision to scrap the sucre for the dollar. They were joined by about 20,000 union members. Salvador Quishpe and fifty demonstrators occupied San Francisco church, vowing to stay until the government renounced its IMF agreement and ended dollarization. [35] The protesters not only opposed the price hikes of early April but also hoped to put pressure on the government to move with moderation in devising further austerity measures, which everyone expected would come within weeks. CONAIE's Blanca Chancosa joined FEINE's Murillo in condemning Noboa's first three months as a disaster. [36]

Labor unions, organized in an alliance called the Popular Front, called a strike on May 15 to push for wage increases and to stop privatization of telecommunications, the hydrocarbons sector, power generation, shipping, airlines, and water purification plants—in all, half of the state's holdings. The National Union of Educators called an indefinite strike on the same day, rejecting the government's offer to raise teachers' basic salary by $2 per month and demanding instead a base income of $100 per month, up from the average of $65. The Union also opposed what it charged was a government plan to privatize education. Some 142,000 walked out, and classes were cancelled for three million students. In Quito, teachers marched on the presidential palace but were pushed back by police firing tear gas. [37]

The teachers' strike was joined on May 18 by some 24,000 hospital workers, who proceeded to attend only to emergency cases in state hospitals. The doctors' demand was a minimum salary of $300 per month. Their average salaries were about $50 per month, and they demanded 50 percent pay hikes. The teachers ended their 46-day walkout after the government agreed to increase their starting pay by 81 percent and promised not to privatize the public school system. On the latter score, Noboa denied ever favoring the much-discussed idea. [38] Health-care workers went back with a 50 percent boost in pay and a government promise to pump an extra $10 million into the public health system.

The protests and strikes of April and May had three fundamental goals: increasing salaries, repealing the price hikes of early April, and putting pressure on the government to move with moderation in devising further austerity measures. New belt-tightening proposals were expected within weeks. Noboa announced gradual austerity measures on May 25 that raised the price of gasoline for vehicles and increased electricity rates. The president simultaneously tried to cushion the impact by raising public-sector salaries and upping social security pensions. As anticipated, in the wake of the new directives, a 48-hour nationwide strike was called for June 15–16 by the leftist-led Patriotic

Front, an umbrella group representing unions, students, and grassroots organizations. They were joined by public school teachers and public health workers who had been on strike for weeks. And while 150 teachers demonstrated in Quito and thousands of students went without classes, businesses and transportation carried on as usual. For a time, tensions rose between the Indian movement and strike leaders. Some said the Indian movement declined to participate simply to show that no one could succeed without it; others claimed CONAIE thought the strike was premature and wanted more time to present its proposals to the government and allow Noboa to respond. The president's principal adviser, Vivas, said the administration was not concerned with the threat of continued strikes and protests: *The majority of Ecuadorians accept that it was necessary to take measures of economic adjustment. I don't say that, the polls do. The protests do not scare us in any way. This country is not going to drown. We are going to do what we have to do in order that it does not.*[39]

Among the loan conditions set by the IMF were increases in the price of government-subsidized gasoline and home-cooking fuel as well as higher electric and telephone rates. Likely all these costs would go up severalfold. Rather than a gradual phase-in of the price hikes, on May 7, Noboa said *the bad news should come all at once*, as should measures to alleviate the shock for the poorest sectors. *They are measures we have to take, as without a stable economy without inflation, there can be no social plan*. But a few days later, he gave himself some room to maneuver: *I am a man who consults and revises his position when I have to revise it. That is my style.*[40] Many voices urged a more cautious approach to belt-tightening. *El Comercio*, for example, took the position that since the $110 million in foreign loans for social programs was insufficient to help the poor, and because money saved by reducing subsidies would be a major burden, they should be eliminated gradually *in order to avoid a hard blow to society*.

Economy Minister Jorge Guzmán said in a mid-May interview that inflation would surpass 74 percent in 2000, but would drop to 30 percent in 2001, 15 percent in 2002, and 5.8 percent in 2003. Guzmán advocated the so-called shock approach and wanted to implement harsh austerity measures all at once.[41] It was soon apparent that administration officials were divided over the nature and extent of the belt-tightening steps. Attacking subsidies had always been politically explosive. For the poor they served as a safety net; and many in the middle class thought that without underwriting, many products would reach international prices, which they feared would be beyond their reach. Mindful of the results of local and regional elections for mayors and provincial prefects held on May 21, which gave victories largely to the center-left, Noboa opted for far more moderate than expected price increases. He adopted higher than anticipated salary raises as well. Guzmán's shock position, which once seemed in accord with the president's, fell out of step. On May 23 the finance minister accordingly resigned and was replaced by economist and former Central Bank director Mauricio Dávalos.

Noboa had carefully calculated political realities when on May 25 he chose a gradualist approach that raised the price of energy but granted salary increases and assistance for those at the bottom. Some $10 million was guaranteed for the Development Fund for Indian Communities. He would not, he said, *drown the poor.*[41.1] Over 97 percent of the country's wheat came from abroad, and to ease the burden of buying bread Noboa lowered its import duty. Salaries for the Armed Forces, Police, teachers, and others in the public sector were raised to a minimum of $120 per month, up from the average worker's income in December 1999 of $53 per month. The president urged Congress to raise the minimum wage in the private sector as well. Nonetheless, the National Institute of Statistics and Census announced in June that the poverty line was $112. Noboa increased social security pensions 40 percent.

While he did not raise the price of gas used in the home, gas for vehicles was upped an average of 66 percent (depending on octane level), bringing the price of "regular" to a little under $1 per gallon. Noboa calculated the hike would bring the state over $922 million per year, about $12 million more than had been promised to the IMF. Electricity rates would rise 4 percent per month until they reached increases of 100 percent for businesses, 60 percent for residential use, and 20 percent for the poor. About 96 percent of urban dwellers and 55 percent of rural people had electricity, a total of around 80 percent of the population. *For the well-being of our economy, such products must be paid for at market value,* said Noboa. *Those who have more, step up and pay, Gentlemen. The era in which Ecuador subsidized others has finished. . . . To avoid higher taxes in less favored sectors, we have excluded a hike in domestic gas prices from our social and economic plan.*[42]

While the state insisted it would still auction off a 51 percent stake in its electricity sector—some eighteen government-owned power generation and distribution companies—it pushed back the date of the sales from August to November. The most recent privatization of state assets was in 1995, when the government sold 50.1 percent of its airline Ecuatoriana to a Brazilian-Ecuadorean consortium. Finally, to stimulate the economy, Noboa proposed the temporary elimination of import duties on agricultural machinery, parts, and accessories, second-hand engines, used vehicles, and electrical appliances. The creation of 45,000 jobs on public works projects was part of the proposal.

Expected opposition came from organized labor. The two-day general strike on June 15–16 failed when CONAIE refused to participate, but nevertheless FUT head Wilson Álvarez called the measures *a blow against the people*. Napoleón Saltos of the CSM charged that Noboa maintained *the same style that in the past did not produce results. Instead, it polarized the breach between the rich and poor*. Nonetheless, Saltos said his movement would not resort to the *accustomed response* but instead would *develop a strategic plan of action to change the economic and political program, stimulate production, satisfy social demands, and broaden democracy to benefit everyone*. CONAIE'S Salvador Quishpe claimed Noboa's price hikes would make transportation and food prices more expensive and feed inflation. *In no way does it benefit the poor, even less the rural sectors*, he said.[43]

Most people had moderate responses to the measures, as the president had hoped, and concluded that his pragmatic approach would buy time while state enterprises were sold, the foreign debt was renegotiated, the taxation system was reformed, amnesty was granted, and the economy began to improve. *I ask only for a few months for economic indices and news from the outside to advise us*, Noboa pled, *that we have passed through the most enormous tunnel in our history*.[44] A Cedatos poll taken a few days after the measures were announced doubtless buoyed the president: 45 percent approved of his performance, up from 34 percent at the beginning of the month, while 64 percent backed the fuel price hikes and the revocation of tariffs. Inflation, meanwhile, seemed to be falling. It rose 5.1 percent in May—only half of April's 10.2 percent increase—but it was still projected to reach between 80 and 100 percent for the year, while growth was anticipated to be near 0.5 percent, if not zero. Twenty-five percent of the inflation was attributed to the devaluation that occurred during the first ten days of January.

The economy improved during Noboa's second 100 days. It grew 2.4 percent in the second quarter as compared to the 0.5 percent registered in the first quarter of the year. Seven of the economy's nine sectors improved: agriculture, manufacturing, electricity, construction, trade, transportation, and domestic services. Activity in the oil, financial, and government services sectors decreased. During the first nine months of the year, investment in private companies increased a dramatic 146 percent.

Ecuador had defaulted on $6 billion in Brady bond debt and $500 million in Eurobonds in late 1999. In early August 2000 a team of Ecuadorean officials, headed by Jorge Gallardo, met in New York with a group of about fifty bondholders and Wall Street analysts in an effort to convince bondholders that reducing the country's debt burden would help rebuild its economy. The nation's bond negotiating team proposed reducing the country's debt burden by 40 percent under a proposal to exchange $6.65 billion of defaulted debt for $3.95 billion in new sovereign bonds. Under the terms of the bond swap, which won the support of the IMF, Ecuador would offer $2.7 billion in 30-year bonds and $1.25 billion in 12-year bonds for the old debt, which included Brady bonds and Eurobonds. J. P. Morgan and Salomon Smith Barney acted as dealer-managers for the exchange.

On April 12, Noboa had called on Congress to grant amnesty to the civilians and soldiers who had participated in ousting Mahuad on January 21–22. The issue was complicated on May 4 when Generals Carlos Calle and José Lascano Yañez testified before a military tribunal and contradicted General Mendoza's version of Mahuad's fall. Until January 21, Calle was chief of staff of the Joint Command and Lascano was chief of staff of Land Forces. On January 21, Mendoza was the head of the Joint Command as well as minister of defense.

The two generals claimed that Mendoza had taken them and every other general in the country by surprise when he joined the Junta of National Salvation. Moreover, contrary to Mendoza's testimony, there was no military strategy to eliminate that Junta by having him become one of its members. Mendoza, they insisted, acted on his own, and with the full collaboration of General Telmo Sandoval. *The constitutional succession*, the pair declared, *was the result of the decisive opposition of the generals and admirals of the Armed Forces* to Mendoza's actions in joining the Junta. Mendoza, claimed Calle, wanted power for himself but met firm military opposition to the idea, and *when he renounced his participation in the triumvirate, he proposed that he be replaced by Paco Moncayo.*

In response, on May 8 former defense minister Gallardo asked that General Sandoval be relieved of his post as head of the Joint Command, arguing that *his presence harms the Armed Forces*. Sandoval was widely perceived as supporting Mendoza's decision to join the short-lived Junta on January 21. Gallardo's call was joined by others when General Villarroel testified next before the tribunal. He swore that on the evening of January 20 and the morning of January 21, the Police were both prepared and eager to dislodge the Indians who surrounded the Congress and Supreme Court buildings, but that General Sandoval asked them to postpone their attempts on three occasions. Villarroel also claimed that the biggest surprise came on the morning of January 21, when, as noted earlier, busloads of soldiers arrived: *At first it was thought they came to reinforce security, but later they joined in capturing Congress.* He never knew who conspired against Mahuad, he told the judges, but admitted hearing rumors of conspiracies. Everyone was well aware of *the public announcements the Indians made in respect to their intention to dissolve the three powers of the state.*

Mahuad's finance minister, Alfredo Arizaga, confirmed on May 8 what Interior Minister Vladimiro Álvarez had testified to earlier, namely, that General Mendoza, on behalf of the high command, had effectively demanded Mahuad's resignation on January 21 when he asked him to make a *democratic decision* in the face of the crisis. The testimony highlighted what had long been obvious: the high command had made little effort to save Mahuad. Colonel Lucio Gutiérrez, perhaps the most interested of those incarcerated over the entire affair, commented: *Those who participated in the acts of January were the highest military officials*; and Antonio Vargas also said, *The generals of the three branches knew what was happening and tried to proclaim themselves dictators.*

Mendoza denied his colleagues' accounts, calling Calle and Lascano *despicable and disloyal.* He reiterated that he had no choice but to confront the divisive and dangerous elements threatening both the military and the public. Indeed, he had sacrificed his career to save from chaos the institution of the Armed Forces and the constitutional order. He never wanted power for himself, he insisted, and from the moment he joined the Junta of National Salvation, he had intended to dissolve it. The accounts of Calle and Lascano, he supposed, were part of a strategy of Mahuad to place *a curtain of smoke* over government corruption and the real reasons for his ouster. *How could it be,* he asked, *that people who were not there can say what they did not see or hear? The high command adopted a strategy, and that will be proven.*[45]

The conflicting versions of events, Defense Minister Hugo Unda admitted, greatly preoccupied the military as well as other sectors of society. He said he favored amnesty for all involved in the January 21–22 events in order to *regain the peace and tranquility of the country so as to be able to initiate its development.*[46] Unda thought amnesty should cover penal sanctions only and not restrict any disciplinary measures that the military might chose to impose on its members. President

Noboa, noting that polls showed 62 percent of the public in favor of amnesty, supported the same approach, and he added: *I do not believe the colonels should return to command their units under any conditions. . . . The country needs amnesty for national reconciliation, and that is the most important thing . . . the country needs a tremendous amount of tranquility, a profound peace in order to be able to advance, transform, and modernize.*[47]

On the heels of the new testimony by Calle, Lascano, Villarroel, and Arizaga, on May 9 the country's top military commanders—General Telmo Sandoval, head of the Joint Command; Vice Admiral Enrique Monteverde, head of the Navy; and General Ricardo Irigoyen of the Air Force—submitted their resignations to the president. Only Army commander General Norton Narváez, widely perceived as not having been involved in Mahuad's ouster, failed to resign. The departures left the nation with thirty-three generals in active service. Some believed it was a step toward ending controversy within the military and repairing its damaged image during a difficult period.

Noboa saw it that way as well. He accepted all of the resignations and named Vice Admiral Miguel Saona Roca to head the Joint Command. General Oswaldo Dominguel Bucheli took command of the Air Force, and Vice Admiral Fernando Donoso Morán of the Navy. Observers noted that the Navy had gained strength, likely because, along with the Air Force, it had not been divided as had the Army and Police during Mahuad's overthrow. Noboa had flown on board a Navy plane from Guayaquil to Quito to assume the presidency on January 22. Moreover, Saona Roca, with the greatest seniority, was a logical choice for the military's top spot.

Although Juan José Pons, the head of Congress, claimed the peace of the republic was at stake, only fifty of the eighty-two votes necessary for amnesty could be found.[48] In favor were DP, ID, Pachakutik, FRA, and MPD representatives. The key was the PRE, which maintained it would support amnesty if Abdalá Bucaram were included. Most members of Congress favored a complete amnesty, as did CONAIE. The PSC remained opposed to amnesty for the colonels and generals, although they favored it for soldiers of lesser rank. Party leader Jaime Nebot explained their position: *It was not impromptu, but rather a coup. We want the process to proceed to its conclusion, so they can be sanctioned.*[49]

The power balance in the legislature altered dramatically on May 29. From his exile in Panama, Abdalá Bucaram changed his position and announced that his PRE would support amnesty. The shift came when public opinion polls showed that 67 percent of the people favored the measure and when the May 21 local and provincial elections resulted in losses for Bucaram's party. The blow bolstered the view that Roldistas needed to appear conciliatory, avoid taking a position on amnesty that only the PSC favored, and ally with the center-left that favored pardons.

Bucaram, with his characteristic flair, announced *the sacrifice of the leader in favor of peace in Ecuador* and denied any hidden motives for the reversal of his position. He explained: *We decided to vote in favor; we are Christians, we know that God's justice will compensate us. It was necessary to turn the other cheek.* Febres Cordero, mayor of Guayaquil and a powerful PSC leader, was certain Bucaram had obtained something for himself in the bargain and assured his fellow countrymen: *You can be sure that there is something, because Bucaram does not give anything for the civic good or for free.*[50] In any event, the additional twenty-two votes gave those in favor of amnesty a total of ninety-two, ten more than the number needed.

Ninety deputies voted on May 31 *to grant a pardon to civilians and members of the Armed Forces, authors of and participants in the actions taken on the 21st and 22d of January.* Twenty-four PSC members voted "no," one member abstained, and one DP deputy was opposed. Febres Cordero was critical of Congress, saying, *He who sows winds will harvest storms.* He warned that *the precedent is terrible, there is no longer a scale of values in this country,* and he lamented that every day more respect for the law was lost.[51] The newly elected mayor of Quito, and one of those pardoned, General Paco Moncayo, saw Congress's vote differently: *I celebrate the amnesty . . . because in such difficult moments for the country it should not maintain such a profound and severe wound.*[52]

When given the news of his own pardon and that of 279 others, Colonel Gutiérrez said: *It will contribute to reducing social and political tensions which are very high at the present time.* He suggested that he himself might enter the political arena *in order to create a political system that seeks social improvement, that combats the traditional one which seeks to benefit only the few.* Amnesty was *a joy, but one moderated because we cannot be happy while more than 80 percent of Ecuadoreans live in poverty, while two million have immigrated abroad in search of better work opportunities, and while corruption remains unresolved.*

The colonel had spent his 132 days in prison writing his version of the January 21 events. Gutiérrez reiterated that he was thinking of *initiating a political career. . . . I would try to form a movement of national identity in which would come together the aspirations of all the people and nationalities of Ecuador. . . . We think of a great movement that integrates the Indians, the blacks, the Mestizos, and all of the underprivileged of this country . . . violence never leads to anything positive. My words and deeds have always been pacifistic. I demonstrated that in 1997 and in 2000. . . . I would love to do the work realized by Venezuelan President Hugo Chávez.*[53] Bucaram used the occasion to put his own spin on events: *I understood that the insurrection of 2000 was an insurrection of the people. It was a kind of protest; it demonstrated that the social, political, and military sectors were indignant at having been used in the fall of Bucaram.*[54]

Despite the amnesty, seventeen officers were sanctioned by the Council of Generals for participating in the January 21 coup. They were placed at the disposal of Admiral Unda for three months, during which time they would be limited to administrative duties, after which they would be relieved of their commands and dismissed from the Armed Forces. Lawyers for the seventeen said they would appeal the decision as contrary to the amnesty given by Congress, but on June 12, eight of them abandoned their careers: Colonels Lucio Gutiérrez, Fausto Cobo, and Francisco Fierro; Lieutenant Colonels Mario Lascano and Víctor Gortaire; Major Víctor Avenal and Captains Sandino Torres and Cabo Germán Carrasco. Gutiérrez explained: *There is a general climate of intranquility in the country and we hope that this decision will overcome the differences, and those within the Armed Forces as well.* Moreover, in late June, over 200 junior officials were sanctioned for their roles in Mahuad's overthrow. Most were held under arrest, incommunicado, for periods ranging from three to nine days.

Cedatos polls showed confidence in the Armed Forces declining from 72 percent in 1996 to 68 percent in February 2000, and dropping still further to 59 percent in May 2000. Confidence in virtually every other institution declined as well, including in the government in general, the courts, Congress, banks, labor unions, the Church, universities, and the communications industry. Defense Minister Unda summed it up by saying: *The Armed Forces are bleeding after a series of events that have meant the exit of a courageous group of generals, colonels, and high officials, especially in the ground forces. The Armed Forces are the only institution that was affected by those events even though they weren't military in character, but political.* Yet the actions were necessary, he insisted, to maintain discipline. *The challenge is now to strengthen the military's democratic consciousness to avoid any political distortions in the future.*[55]

The need for a united military surfaced at the same time. Colombia's leading Marxist rebel group, the 17,000-strong Revolutionary Armed Forces of Colombia, or FARC, was Latin America's largest surviving 1960s rebel army, which some said controlled as much as 40 percent of the country in 2000. Colombia's internal conflicts had taken more than 35,000 lives during the 1990s, and in early May 2000 FARC said its fighters would seek sanctuary in Ecuador and Peru if forced into retreat. (Ecuador and Colombia share a 370-mile border.) A major push by Colombian security forces into the south threatened to have a huge impact on the border area. It was encouraged by billions of dollars in U.S. military aid and strategy, Plan Colombia, to support President Andrés Pastrana's effort to revive the ailing economy and to destroy guerrillas, drug laboratories, and cocaine plantations in the jungle-covered regions under FARC's control.

An Ecuadorean Army patrol on May 15 intercepted rebels belonging to the Revolutionary Armed Forces of Ecuador (FARE) in Sucumbios province near the border with Peru. FARE was widely believed to be a dissident faction of FARC, although little was known of the Ecuadorean group. Two rebel members were killed. The Army continued its regular patrols along the Colombian border, and the Air Force made regular flights to determine if the rebels had infiltrated Ecuador. The United States had given Ecuador $89 million to combat narco-traffickers intimately tied to the guerrilla movements in Colombia. As fears heightened that FARC would spill into Sucumbíos province as a result of being pushed south, Chief of Staff Saona Roca and Foreign Minister Francisco Carrión disagreed over whether Ecuador should be included in Plan Colombia. Carrión believed it should, while Saona did not. According to Saona, *Plan Colombia is a U.S. plan that applies to Colombian territory, whose objective is to stop the guerrillas and the narcotics traffic. That does not involve Ecuador; we do not participate.*[56] Bishop Gonzalo López of Nueva Loja had another view: *This damned Plan Colombia is a plan for the annihilation of poor Colombian peasants. But Ecuadoreans are by nature pacifists, I would say even passive*. Luis Villacis, head of Ecuador's Popular Front, agreed: *Manta will be used to attack Colombian guerrillas and peasants—acts of war that we cannot allow.*

Nonetheless, the fallout from Plan Colombia was increasingly felt in Ecuador. The United Nations predicted some 20,000 to 30,000 refugees, driven by the crackdown on the guerrillas and the spraying of drug crops with herbicides, would soon cross the border from Colombia. Noboa's government planned to build refugee camps to control the influx. At the same time a plan for social and economic reactivation of the northeastern area of the country, centering on Sucumbíos province, was under discussion to prevent what was occurring in Colombia from happening in Ecuador. A special military unit, moreover, was being trained to keep armed groups from crossing the border. Noboa said: *All of Latin America believes that it is necessary to get rid of drug traffickers. We will work to protect our borders, and if it happens that the war comes across the border or that the coca farmers move their operations to Ecuador, the U.S. has promised to help.*[57] Grave concerns were voiced about the U.S. Air Force presence in Manta, about whether, despite assurances to the contrary, Ecuador would be drawn into what most saw as a U.S.-Colombian problem that should not involve them. Peru's President Alberto Fujimori warned that the U.S.-backed anti-drug initiative in Colombia could escalate guerrilla activity to the point of threatening regional stability, since the leftist rebels, who financed their insurgencies with drug money, were likely to be pushed across Colombia's border into northern Peru and Ecuador.

In June 2000, Ecuador's Banking Superintendent Juan Falconi issued his report on investigations against Filanbanco, perhaps the most notorious of the financial institutions accused of corruption during Jamil Mahuad's administration. Falconi found that the bank improperly used millions of dollars lent to it by the government to help it avoid bankruptcy. Contrary to conditions attached to the loan, Filibanco managers transferred almost $100 million from their headquarters in Guayaquil to offshore operations just before the bank was finally taken over by the AGD. Many of the offshore operations, moreover, reported fake records for nonexisting transactions and made substantial loans without adequate collateral and guarantees. After the release of Falconi's report, attacks increased against the chief government attorney, Ana Mariana Yépez, who had been investigating the case for fourteen months. She was harshly condemned for trying to cover up massive corruption in Filanbanco and for neglecting her duties as a prosecutor.

Corruption charges were brought against Mahuad in midyear as Congress authorized the Supreme Court to prosecute him. The legislators decided that the Court did not need congressional authorization to charge and prosecute former presidents; they only needed it to go after sitting ones. The action cleared the way for pursuing several criminal cases, including those against Abdalá Bucaram as well as Mahuad.

Local elections had been held on May 21. The two biggest losers were the parties of the recently deposed presidents, Bucaram's PRE and Mahuad's DP. In regional terms, the elections demonstrated that ID and DP were the strongest parties in the highlands and that a coalition of ID, Pachakutik and New Country outweighed all other alliances. On the coast, the PSC and the PRE were the strongest. In ethnic terms, the Indian vote was massive, resulting in unprecedented victories for Pachakutik on the local level. Analysts forecast a PSC-ID presidential contest, and many speculated that Pachakutik would back ID, as they did in many local contests for mayor.[58]

The PRE suffered overall losses, and to make sure they did not happen again Bucaram asked about three hundred party delegates to Panama to dissect the election results and discuss the party's position on pending legislation and other issues. Former president Mahuad, meanwhile, gave seminars and speeches in the United States and Europe.

In mid-July, Noboa sent a controversial emergency bill to Congress, for action within thirty days, to allow private companies to invest in joint ventures with the state oil company, Petroecuador. They were not to buy shares in the enterprise itself but they could take up to 51 percent of the shares in electricity and telecommunications companies. *This is a bill that will permit us to attract investment; it does not mean privatization at any cost*, said Ricardo Noboa, the president's brother and his top privatization official.[59] The president, he added, believed that economic reforms could not drag on much longer without destroying the nation's credibility and undermining the apparent success it had in recent foreign debt and loan negotiations. In the 1970s, Ecuador and most oil-exporting nations created state oil companies and began the process of pushing out foreign producers. By the end of the century, Ecuador and most others were trying to lure them back in. Oil reserves had declined, and an intensive search was on for finding new wells. Some adapted to the new economic reality, while others clung to nationalist and redistribution concerns.

CONAIE and union opposition was immediate. Among other things, the Indian movement feared that privatization of natural water sources would impose a financial hardship on peasants dependent on free access to irrigation. It was also concerned that privatization of oil, communications, and some eighteen electrical power companies would lead to price hikes and uneven distribution of services and resources. CONAIE and other groups claimed that the sales would give away significant national assets to foreign companies, which could then exploit them at will, and that there would be virtual giveaways of public property to vested interests. A handful of powerful individuals, they claimed, had in the last year purchased certificates of deposit from local banks that they then could redeem at face value to buy government enterprises and utilities.

The Law to Promote Investment and Citizen Participation was dubbed Ley Trole II. The president promulgated it on August 18 after Congress, divided into two factions (each of which refused to recognize the other's legitimacy) failed to take action on the bill within thirty days as required by the constitution. According to the constitution, every two years Congress's head must come from the party with the second largest number of seats, rotating the post with the party with the most seats. The problem was how to interpret the constitution.

After August 1, each of the two factions had its own elected leader and claimed to make up the legitimate Congress. One group of legislators met on the first floor of the building, while the other convened on the second. The majority of center-left parties opposed Ley Trole II and chose three-term congresswoman Susana González as head. González, from Cuenca, was the first female president of Congress. She stated: *A connection exists between political directors and groups with economic power, who try to control what is done in Congress. The fact that certain articles in some laws are approved that benefit certain economic groups is something very negative and has to be combated.*[60]

The contending faction, led by the conservative PSC, elected Xavier Neira as its leader and supported Ley Trole II. While personal rivalries were at play, with Neira disliked by many, the division in Congress reflected a power struggle for who would control the body and set its policy agenda. The center-right had provided most of Noboa's support; the center-left had resisted much of his economic program. PRE congressman Víctor Hugo Sicouret put it succinctly: The struggle *is over political domination of the country. And in that regard, the fight has just begun.*[61]

González resigned from the PSC two weeks before her election when she learned that party bosses Febres Cordero and Nebot were backing Neira for the post and were demanding that she and others sign a refusal to be a candidate. González and her backers argued that pursuant to the constitution the legislature could select anybody elected to Congress on the PSC ticket. Neira and his supporters countered that the head of Congress had to be a current party member. The 1998 constitution specified only that the legislature had to be led until 2002 by a member of the second largest party, the PSC. Neira argued that González's selection was unconstitutional. The party entitled to name the head of Congress, most PSC members claimed, should select that candidate.

Many analysts saw power politics at play far more than constitutional interpretation. They noted that the national charter, the country's seventeenth, had hardly been followed in either letter or spirit in recent years. The military had forced out both Bucaram and Mahuad; while Arteaga was prohibited from succeeding Bucaram in February 1997, Noboa was allowed to replace Mahuad in January 2000. Bucaram had been removed from office by Congress on the dubious grounds of mental incompetence, and Arteaga had just as valid a right to accede to the presidency as had Noboa.

A constitutional court agreed with Neira, as did President Noboa, who said as early as July 25 he thought the constitution gave the post to Neira. González stepped down. On August 28 a new legislative chief was chosen, Hugo Quevedo of the PSC. He had been well liked as a deputy for over a decade by Bucaram and was on good terms with the PRE. Quevedo had been a member of the PSC for only two years and considered himself politically flexible. He was elected by the same center-left coalition that had chosen González. Because Febres Cordero refused to back him, Quevedo lacked the support of PSC deputies.

Most legislators wanted to modify the privatization bill. In an effort to mollify his critics, who accused him of heavy-handed and unconstitutional conduct, President Noboa promised to introduce another Economic Transformation Law to amend Ley Trole II, which he nonetheless put into effect. Before promulgating the new law, the president and the head of the Armed Forces addressed the graduating class of the Escuela Militar Eloy Alfaro on August 10. Both spoke to the constitutional crisis in Congress. According to Noboa, the division of Congress into two camps—each claiming to be the only legitimate one—had once again made the nation subject to international ridicule.

In sum, during his second 100 days, Noboa chose moderate as opposed to severe austerity measures, and despite some dissent most people accepted them as necessary. Although he continued to face strikes and protests, none caused severe instability. Major shifts within the military were made, amnesty for those who ousted Mahuad was granted, and the colonels instrumental in the overthrow were separated from the Armed Forces. Moderates made gains in the Indian movement, and in the nation's politics the left won in the highlands and the right on the coast, albeit in local elections. While neighboring Colombia's guerrilla movement showed signs of spilling into Ecuador, for the moment, at least, the threat seemed under control.

CONAIE's Antonio Vargas summarized Noboa's policies by saying: *In six months this government has not made a single change to benefit the country, and what it wants is to sell Ecuador to private enterprises.*[62] While not concurring with his critic, Noboa offered a partial explanation for his nation's plight. Globalization and international lenders, he said, left Ecuador and other poor countries with *increasingly less margin to apply policies that mitigate against unemployment and inequality. International debt consumes more than 50 percent of the national budget in some countries, which does not permit the execution of programs to attend to the needs of the poorest of the population . . . alarming levels of poverty and misery in the great majority of countries are increasingly caused by forces outside of national frontiers and beyond national control.*[63]

Conclusion

Two weeks after the fall of Jamil Mahuad, *El Comercio* editorialized that few if any of the participants in the January 21–22 events wanted to call it a coup: *There was no coup d'état, they say, and they accuse former [Foreign] Minister [Benjamín] Ortiz and former president Mahuad of having wanted to give themselves a coup d'état. . . . Not the Indians, not the colonels, not the generals, nobody accepts that a coup d'état took place on Friday, January 21.*[1] Jacinto Velásquez, the defense attorney for the junior officers arrested for their roles in taking over Congress and the Supreme Court, also asked who did what. He stressed the fact that the junior officers never used firearms and never intended to; they simply stepped aside and let people pass through the doors of Congress at a time when polls showed that 90 percent of the nation wanted the president out. On the other hand, Abdalá Bucaram called his ouster in 1997 a coup and gave that name to his account of his fall, *Golpe de Estado*, although he noted that most congressmen refused to call it that. The same seemed true in 2000. If what happened on January 21 was a coup, then, despite denials, it was not a traditional one.

The event's prime movers did not hesitate to place blame or to give credit. Mahuad, Carlos Mendoza, José Gallardo, Antonio Vargas, and Gustavo Noboa all had their own versions. Generals Mendoza and Telmo Sandoval and other top military officers all wanted to get rid of Mahuad, as did the majority of soldiers, politicians, and other citizens. The officers allowed CONAIE to form a human double ring around the Congress and Court buildings to force the president to depart, a decision that carried consequences more complicated than the generals intended. They were caught by surprise when Colonel Lucio Gutiérrez and a cadre of junior officers made the intolerable move of joining the Indians. The only way out of this volatile situation was Mahuad's ouster, the dissolution of the Junta that divided the nation, Noboa's assumption of power, and the separation from the military of numerous junior and senior officers in the interests of both institutional and national unity. Others offered their own interpretations.

Deposed former president Mahuad, for example, blamed the military, but like others he was uncertain about what to call the event, as indicated in his televised address to the nation on January 21. *If this is about staging a military coup and taking power by force, Gentlemen, then take power by force.* The senior officers of the Joint Command had demanded his resignation, and he saw complicity between them and the lower-ranking officers who had treasonously joined the Indian protesters. The military failed in its obligation to keep order, protect the buildings and institutions of government, and preserve the constitutional order. Betrayed, Mahuad complained of having been *treated worse than a dog. The great problem of this country,* he concluded, *is the lack of unity and the lack of even minimal agreement within the political class.*[2]

Mahuad's interior minister, Vladimiro Álvarez, declared that the government was toppled by *an important ethnic minority of a maximum of 15 percent of the population* as well as by junior officers who *publicly rejected the highest level of military command* and by generals who refused to obey orders. Also culpable were civilians who believed *it was better to continue living in a supposed bonanza of consumption that had characterized Ecuador in recent years based on uncontrolled public spending, the destruction of natural resources, and public and private debt,* and who refused to accept the need for *cutting public spending and raising taxes.* Mahuad's abilities were not the problem. According to Álvarez, *The central question is not whether a decision should have been made a month before or a month later . . . that is a very poor interpretation of events. . . . That is not the central problem of Ecuador; the problem of the people is that everyone looks after his own interest and is incapable of thinking of the country.* Furthermore, rather than follow article 109 of the constitution's requirements for revoking presidential power, including a popular referendum, a minority substituted themselves for the law and left Ecuador's democracy in shambles. The country and the Armed Forces forgot, Álvarez said, that *law is not an expression of force; force should serve to protect the law.*[3]

General Gallardo, Mahuad's defense minister until ten days before his overthrow, blamed Mendoza and Sandoval, claiming the generals used idealistic and frustrated junior officers as well as the Indian protesters to try to take power for themselves. Both the heads of the Joint Command and the leader of the Army were thwarted because others in the high command, and no less than twenty regional military heads, rejected their grab for power.[4]

General Mendoza denied complicity in the initial events that presented high-ranking officers with the choice of military fracture and bloodshed, on the one hand, or Mahuad's departure, on the other. Mendoza claimed he and other generals and admirals were forced to oust the president to preserve the unity of the Armed Forces and to avert civil war. They never sought power for themselves, he swore. In his view, *the Ecuadorean people had revoked the mandate they had conferred on Jamil Mahuad, because of his incapacity to govern, ignorance of public administration, subordinating himself to the financial interests of some bankers, not complying with his campaign promises, and being without a plan for governing.*[5]

Former general turned congressman Paco Moncayo, who had played a crucial role in Bucaram's exit a few years earlier, did not believe the junior officers or anyone else had carefully planned Mahuad's fall. *The colonels were worked up about corruption,* he said. *They did not have a leader or a visible plan. It was nothing but an explosion against a terrible circumstance.* He thought the events were to be expected. *When military leaders do not stand up to eliminate the corruption that the country is seeing under this government, officials appear who look for a way out, as occurred in Venezuela when Hugo Chávez took a stand. Ecuador has no reason to be different.*[6]

CONAIE's Antonio Vargas gave the Indian movement the bulk of the credit for deposing Mahuad, but added: *We never had any desire to take power. We knew that if we tried to do so, there would be a lot of bloodshed.* He put the events in a decade-long perspective: *The deeds of January 21 did not come only from the year before, but from the great mobilizations we initiated after 1990, from the first Indian uprising until January 21, 2000.*[7]

Colonel Gutiérrez also put January 21 in a broad context: *What happened in Ecuador is similar to what is happening in the rest of Latin America, caused by the neoliberal system that has been completely hurtful and prejudicial to the interests of every one of our countries. Enemy number one of democracy is not the military, the social movements, or the people. Enemy number one is corrupt politicians, those who destabilize our country. They are the ungovernable, not the people. . . . I think that very few countries in the world would support what the Ecuadorean people have supported, so much injustice.*[8]

From exile in Panama came former president Abdalá Bucaram's reaction: *Seeing Mahuad condemn the coup today, and seeing him talk of democracy, gave me the impression of seeing Satan say Mass. . . . The evil 1997 coup prostituted Ecuadorean politics. . . . The last straw was obviously the dollarization of the economy.*[9]

Newly installed President Noboa said it was all an embarrassment and, as we know, labeled the events *buffoonery*—an understandable explanation from a man in the daunting position of having to pick up the pieces and govern. He also intimated that since Mahuad announced on January 21 that he would not stand in Noboa's way, Mahuad in effect gave up without a struggle. Several months later he would tell the press that Mahuad both got into trouble and fell because of inaction.[10]

An analyst who asked not to be identified gave another explanation:

> *Ecuador can be seen as a large hacienda with a handful of managers at the top and the vast majority of laborers at the bottom. In good times the managers give a bit more to the laborers—more pay, education, health care, and so forth. In bad times they take away from the laborers as they can. If the managers have difficulty during good times or bad, they run to the government for help and almost always get it. When the laborers, particularly the Indians, appeal to the government, the managers are horrified and ask, How dare they? Who do they think they are? and label them irresponsible and worse. It stems in part from greed, in part from part racism, in part from cultural formation, and from a mixture of much more inherited from many centuries. Through CONAIE and other movements, the Indian people are now organized to say no to this arrangement. They said no on January 21. It is too soon to tell if the managers have learned anything. It is also too early to know whether or not the Indians can transform the country. But in the last several decades they have made enormous advances in terms of self-pride, organization, and assertiveness, which benefit us all.*

The accounts of Mahuad's fall given by those who participated in the event reflect the complexity of what transpired. Explaining Mahuad's removal with traditional terms and familiar concepts presents a challenge. Was it a coup? A countercoup? An effective popular protest movement? The abandonment of office? Was it an entirely domestic affair? Or did foreign influences dictate the outcome?

There is value in noting the deep divisions within the military between radical junior officers and more conservative senior ones, and in appreciating that the high command acted to salvage as much institutional unity as possible. The well-organized indigenous movement had gained considerable strength and experience, and in January it used its power more effectively than ever before in its decade-long history of national protests and demonstrations. It is equally apparent that Mahuad's support had plummeted, that his chances of long-term survival were remote at best, and that his weakness was reflected in his decision not to fight back beyond his few meager attempts. His low level of support was rooted in the country's dismal economic condition, in its unwillingness to embrace his austerity program any more than it had Bucaram's three years earlier, and in the widely held perception that he was corrupt, weak, and indecisive. Many believed Mahuad simply lacked sufficient leadership ability to take Ecuador out of its worst economic crisis since the Great Depression of the 1930s. Others argued he was trapped by institutional divisiveness, public opposition to belt-tightening, lenders' demands that austerity measures accompany loans, and his adherence to the constitution.

All of these observations contain a measure of truth and beg the question of how each crisis arose. Mahuad's problems were much the same as those that had plagued his predecessor, Bucaram, and suggest that forces far greater than personality and style stood at the core of the country's predicament. The two leaders fell only three years apart. A contracting economy, unpopular austerity measures, widespread corruption, political fragmentation, and massive strikes and demonstrations burdened the animated and colorful "El Loco" Bucaram as well as the calm, dignified, Harvard-educated Mahuad. All these difficulties were decades old.

An analysis of Ecuador's economic and political development since the oil era was launched in the early 1970s explains much of what happened to the two leaders. For the last three decades of the twentieth century, oil was the economy's axis: it buoyed government, industry, and commerce, strengthened organized labor, and swelled the ranks of the bureaucracy and middle class. Revenue derived from black gold was central to raising new expectations and demands for socioeconomic and political change and to fomenting organizations determined to advance their own interests. Although the transformation from 1970 to century's end was substantial, the process of modernization was far from complete. The reliance on petroleum made the economy lopsided, dangerously dependent on foreign markets, and extremely sensitive to forces it felt powerless to influence. One administration after another spent and borrowed at unprecedented rates, only to be left with huge debts when the price of crude oil eventually plunged. Austerity measures imposed after the initial bonanza to keep expenditures in line with revenues provoked opposition from an increasingly more inclusive populace accustomed to low taxes and high subsidies—luxuries that were possible when oil prices were high but were prohibitive when they fell.

The discovery of oil in the Oriente in 1967 occasioned profound cultural and environmental dislocations in the region, which generated discontent followed by effective protests on the part of the Amazon's organized indigenous population. Two Indian movements came together nineteen years later in 1986 to form the backbone of

CONAIE: ECUARUNARI, established in 1972 in the highlands; and CONFENIAE, formed in 1980 in the Oriente. Each had its antecedents of community- or nationality-based local groups. Ancient struggles for land and culture in the highlands, and the more recent battles for the same against petroleum incursions in the Oriente, finally united the vast majority of Indians in the two disparate regions. CONAIE was a major force to be reckoned with by 2000. For many Ecuadoreans, the extent to which government oil revenues were utilized to promote social and economic welfare left much to be desired; and for the Indians more than for most, oil wreacked cultural and environmental havoc.

Expectations that the economic and social changes driven by the oil boom would transform traditional political culture were unfulfilled. Customary aspects of civilian politics, including regionalism and personal allegiances, were reflected in the proliferation of new parties and interest groups. Already antagonistic interests both generated and sustained political fragmentation. Ecuador was unable to reconcile the gap between the few who were rich and the many who were poor or to balance the redistribution of income with economic dynamism, growth, and entrepreneurship. Endemic poverty and inequality, as old as the country itself, continued to thwart democratic stability and economic prosperity.

Oil and its consequences also altered the orientation and conduct of the Armed Forces. In an interview several months after Mahuad's fall, General Moncayo reflected that the military began its active participation in the economy in the early 1970s when the petroleum era began; it even developed its own enterprises. By century's end, however, the institution had a far more community-based approach. Recalling the 1970s, Moncayo said: *That was positive because it was in defense of the national interest. Now it almost does not exist, although we have a mixed economy. It was born in the epoch of import substitution, as an assist to industrial development*. The general saw a far different armed institution by 2000: *Now the military assists in human development, works in poor districts, works with the Indians. The military helped the Indians. That union has been positive. Thanks to that union we have had social peace.*[11]

Mahuad's interior minister Álvarez, offered a different perspective. He believed that with *the first barrel of petroleum. . . . Ecuadoreans, as a country, began to live in a fantasy*, with subsidies for domestic and industrial gas, transportation, telephones, electricity, and more. *Great projects . . . lines of credit . . . tax exemptions. . . and subsidies* were pursued with a lack of financial responsibility. By 1998, 70 percent of Ecuadoreans were poor and 45 percent were indigent. *Among the first decisions of Jamil Mahuad*, he contended, was to *leave that fantasy initiated in 1972 behind and realistically face its economic and social reality*. That decision led to social upheaval and to the president's downfall.[12]

Comparisons between Bucaram's fall in 1997 and Mahuad's in 2000 reveal both continuity and contrasts in terms of prime movers against the regimes, the tactics employed to oust the president, respect shown for the constitution, the forces choosing successors, the selection of successors, the role of the military, and the U.S. government's part in the fall of each man. On the other hand, mostly continuity characterized the challenges faced and the policies pursued by all governments from 1996 to 2000.

Labor led the charge against Bucaram, and the unions were immediately backed by the vast majority of citizens, organizations, and interest groups from all strata of society. The Indian movement led the move against Mahuad and was joined by virtually everyone else. Nationwide strikes and massive street demonstrations were common tactics employed against both executives, and for weeks before each overthrow several former presidents urged the nation's leader to step down. During the 1990s there was a new and powerful force on the scene—the Indian movement—in addition to the older labor unions, business associations, and student federations. CONAIE could mobilize the indigenous population and put enormous pressure on the Pan American Highway as well as on the streets, plazas, and public buildings of towns and cities throughout the country. CONAIE both preached and practiced nonviolence, and its tactics and principles gained it widespread public respect. Its Levantamiento of January 2000 was bloodless, as were its previous massive demonstrations, and the organization was central in tossing out two presidents. Within less than a decade and a half of its founding, CONAIE was close to center stage. And though checked by the overwhelming power of a sufficiently united military, it stood poised to give indigenous people, blacks, and other long-maligned and excluded citizens a greater voice.

A common rallying cry for the disparate groups who joined to topple two presidents was the vexing issue of corruption. An endemic national problem for centuries, it had doubtless been generated by poverty, low government salaries, and the force of tradition. Yet the ancient plague took on greater significance in times of economic decline, austerity, and sacrifice. It haunted Bucaram, Fabián Alarcón, and Mahuad alike, just as it had tormented their predecessors. Corruption took multiple forms, including customs theft and financial favors to and from bankers, but the impact was the same: most politicians were perceived as tainted, and their legitimacy as well as that of the entire governmental structure was severely weakened. While the U.S. government forced Ecuador to pursue austerity measures by withholding loans and threatened political and economic isolation if democratic processes were not followed, similar pressures were not applied to root out corruption. It was too deeply embedded in the country's democratic political system and free-market economy to be significantly reduced in the short run. While U.S. government officials condemned corruption and backed OAS programs developed to reduce it, they did little else.

Some presidents and politicians stole more than others; of course, some did not steal at all. Leaders often accepted the corrupt practices of others as the price for garnering the political support they needed to enact their programs. Allegations of stealing followed by arrest warrants, well founded or not, were effective in keeping ousted politicians abroad, as both Bucaram and Mahuad can attest. Many analysts saw corruption as a culture problem, rooted in centuries-old habits and expectations that would not change overnight.

With the old regime out, the major question in February 1997 and January 2000 was who would assume power. In 1997, Congress and the vice president contended for authority; in 2000 the legislature was pushed aside and the choice for Mahuad's successor was between the coalition of Indians and junior military officers and the vice president. Constitutional concerns came into play during each ouster, and soldiers and politicians felt pushed by the U.S. embassy to follow the legal dictates and entrust the vice president with power. In 1997, Congress was a major actor both before and after Bucaram's removal. It decided how to depose the president and extended that role to picking his successor, although it paid a measure of homage to the constitution by having the vice president take command briefly before its own choice took over. In 2000, with Congress and the Supreme Court shifted to the sidelines, and with greater pressure exerted by the U.S. embassy to elevate the vice president, political succession was faster and far less complicated.

One of the Indian movement's many frustrations was that the bulk of the land was controlled by so few people. An even smaller group, the military officers, traditionally played an equally disproportionate political role. Their involvement began with the wars for independence, as the soldiers who created the country insisted on ruling it. Those in uniform saw their role as filling a political vacuum until responsible civilian parties emerged. But even after the rise of fragile political parties, usually centered around an individual, politics remained largely a battle between competing factions of the elite, many of whom appealed to the Armed Forces to gain or hold power. The military intermittently continued to fill power vacuums or provide stability in times of crisis.

For roughly two decades before the events of February 1997 and January 2000, soldiers championed constitutional government and resisted temptations to take power. But as Ecuador seemed out of control at century's end and plunged deeper into economic and political chaos, the men in uniform did not hesitate to return to the fore, if only to engage in quick interventions. In 2000 it appeared that the Armed Forces were torn in several directions. The military was transformed during the century's last three decades, and like the indigenous movement it was moved by outside forces, including the example of president and reformer Colonel Hugo Chávez in nearby Venezuela. Another influence was the world's only economic, political, and military superpower, the United States.

The military was united when Bucaram was felled; when Mahuad was toppled, it was divided between senior and junior officers. In 1997, Armed Forces cohesion made it easier for Bucaram to maneuver and combat his opponents, and unity made it easier for those in uniform to sit on the sidelines longer. In 2000 a divided military was forced to take prompt action to protect against deeper institutional divisions and possible armed confrontations. There was little time to take politicians into account. Because of institutional unity in 1997, there were no military shakeups, resignations, or arrests accompanying the change of government. The divisions in 2000, however, led to major shifts in top positions as well as early retirements, arrests, and expulsions. Some military officials who took prominent roles during the fall of each president later sought political posts. General Moncayo, for example, played a crucial role in Bucaram's departure and in the selection of his replacement. He retired from the military, became a congressman in 1998, and in 2000 backed the movement of Indians and junior officers that ended Mahuad's regime. Later that year he was elected mayor of Quito. Colonel Gutiérrez, a key player in Mahuad's fall, left the military and launched a bid for the nation's highest office.

The U.S. government and the other thirty-three members of the OAS were committed to strengthening democracy and punishing countries that deviate from it. In 1991 member nations authorized the OAS to take any action it deemed appropriate to restore democratic normalcy where it is interrupted, and a 1995 agreement held that if there is a suspension of democracy, there is a suspension of OAS membership. U.S. State Department officials had frequently said that additional pressures were warranted. Political stability and Armed Forces unity in Ecuador were important to Washington for more than ideological reasons, as political upheaval and military division could only weaken the country's response to the guerrilla and drug cartel threats from Colombia. Economic interests were also at stake; the United States was committed to free markets and neoliberal policies throughout the hemisphere, as stable economies, liberal trade, and investment would benefit its own economy. The United States did not topple either Bucaram or Mahuad. In the case of the former, however, the American ambassador provided the military with data on corruption that animated the already ample opposition. It was clear to soldiers and politicians alike that Washington would make no efforts to save Bucaram so long as at least an outward show of legality was maintained in his removal and replacement. The same can be said of the OAS. And so Rosalía Arteaga went through the charade of assuming the presidency, only to resign and let Congress again elect Alarcón to the top spot. The U.S. role in Mahuad's removal was also minimal, but as in Bucaram's fall its stance was of great importance in determining how his replacement was picked. U.S. officials announced that essential foreign loans, aid, and investments would wither unless constitutional procedures were followed. Thus, the vice president would succeed the president for the second time in three years.

In the weeks following Ecuador's crisis in January 2000, the United States and the OAS moved quickly to prevent another breakdown. American leverage was enormous, as it was able to thwart loan restructuring and new advances of money as well as to proceed with more of each. It had used that influence for decades to help shape and mold public policy and political conduct in Ecuador and elsewhere in the region. In return for favorable loan treatment, financial reform was essential for Mahuad's successor, just as it had been for him, which meant imposing a good measure of financial austerity. But now there was an increase in aid to Indian communities, given both to meet urgent needs and in the hope that the money would help moderate indigenous demands. Anticorruption efforts were deemed necessary as well by the U.S. and multilateral lenders, and the military was to stay in its barracks. The basic outline was far from new; its impact remains to be seen. What was clear was that for thirty years, outside forces played major roles in Ecuador's evolution—from oil companies, missionaries, environmentalists, and human rights activists to U.S. ambassadors and international lending agencies.

The backdrops to the two presidential overthrows consisted of corruption, rising prices resulting from austerity measures, proposals to make major changes in money policy (convertibility and dollarization), overdependence on fluctuating international oil prices, and a worsening economy. Common policy objectives of the two presidents were deficit reduction, reducing government subsidies, privatizing state enterprises, and winning international loans. Political fragmentation and deep divisions within Congress stymied the efforts of both men to implement their economic programs. A striking difference between the two administrations was that while Bucaram campaigned on a populist platform, once in office he adopted an austerity program, whereas Mahuad urged higher taxes and lower subsidies during the election campaign and then as president followed through on both promises.

CONAIE's Antonio Vargas said the Levantamiento of January 2000 was the culmination of many mass mobilizations of the 1990s: *Before, the Indian presence in Ecuador was difficult; it was minimal, it was not recognized on the political agenda of Ecuadoreans. We were considered campesinos, period. Now one sees the Indian movement within national politics. . . . Our advance since 1990 has been very rapid.*[13] CONAIE had forged the strongest Indian organization in Latin America. Ethnicity had always mattered in Ecuador, but after 500 years of exclusion indigenous people were a major force to be reckoned with by soldiers and politicians. Now they had gotten rid of a president and, after Mahuad's fall, had moderated the austerity measures of his successor. Ethnic identity and interests united CONAIE, although as 2000 unfolded there were indications that the movement's moderate and radical forces were growing further apart; the radicals, willing to join with urban labor unions and others, were more determined to challenge unpopular austerity policies with mobilizations and confrontations.

Bucaram, Mahuad, and Noboa governed after CONAIE had joined the ranks of the nation's well-organized groups and demanded payment of the social debt to the poor, particularly to the indigenous population. Old political players, including labor unions, Chambers of Commerce and Industry, student groups, and an array of middle-class organizations, made common cause with CONAIE to the point of sacking Bucaram and Mahuad. All were opposed to lowering the subsidies to which they had become accustomed. Government underwriting for food, transportation, and other staples had been made possible by oil, which accelerated modernization and heavy spending, realigned social classes and political parties, and stimulated Indian organization in the Oriente. While old and new forces joined to topple Bucaram and Mahuad, both times traditional groups triumphed in the end and offered solutions to the nation's problems strikingly similar to those championed by their predecessors. Oil dependency, foreign debt, domestic deficits, and reliance on international lenders were enduring difficulties; efforts to boost oil production, cut spending, and satisfy the demands of the IMF were the continuing responses.

While the future is hard to forecast, it is far less difficult to phrase many of the vital questions facing the nation. What will be the response of the military and others if the economy continues its poor performance, if corruption and poverty advance unabated, if military pay and the institution's budget shrink further? And what will be the reaction of the Armed Forces and others should powerful indigenous protests continue? Will belt-tightening and anticorruption measures be sufficient to keep international lenders parceling out new loans? If so, will the country's old loans be written off? Will they be tied to reform? If so, how tightly? Will the new loans of over $2 billion prove adequate to meet the demands of CONAIE, the CSM, and the general public? Will a pluralistic, multicultural community be established and accepted? Will unity be achieved through diversity or through the fusion of two vastly different worlds and cultures? Will modernization of the state and economy take hold? If so, what kind of culture and community will it help mold? And for whose benefit and expense? Will the country's options continue to depend on powerful international lending agencies and the policies of U.S. governments? Will liberalization of the economy continue along the lines of the so-called neoliberal economic model? If so, will it produce more and distribute better? And for whose profit and disadvantage? Will political stability and inclusive democracy move from theory to reality? Will the freedom and dignity of every Ecuadorean, in particular of the humble and the oppressed, become more of a reality rather than an oft-repeated sentiment?

The answers to these and countless other questions remain to be seen. Historians usually confine themselves to clarifying the past, putting the present in context, and trying to make future developments, whatever they may be, somewhat more intelligible. Many of this work's interpretations and conclusions will be revised or even discarded by this writer or by future investigators. In the meantime, if they serve to generate additional questions, investigations, and analyses, they will have more than met their purpose.

Several broad trends, however, seem likely to remain constant. Ecuador's efforts to reshape its society and to create its own history will not be made entirely under conditions of its own making. The country will continue to face severe restraints as well as opportunities imposed by more powerful nations and by potent economic forces centered abroad. The nation's politics, economics, foreign relations, finance, trade, culture, and more, will be understood only by grasping how everything depends on everything else, how each of these practices is contingent on all the others and on their transformations. Finally, despite broad and deep cynicism and corruption, Ecuadoreans will persist with the hope that their future is under no obligation to mimic their past. They will continue to contest conditions of poverty that effect at least 70 percent of the people, to dispute the nation's grotesquely skewed distribution of wealth, and to challenge arrangements of power and participation which are too closely correlated to skin pigmentation, culture, and ethnic identity. New ways of diverse cultures to know and deal with each other will be conceived and reconceived, and new ways of life will be defined and developed. History is what one makes of it. It has never stopped changing.

Notes

Introduction

1. *Reuters*, Quito, 01-22-00.
2. *New York Times*, Larry Rohter, 1-23-00.
3. Ibid.

Chapter 1, The Land and the People

1. *El Comercio* (Quito) 05-07-00.
2. Author's interview with Galo Plaza Lasso in the spring of 1979 in Quito, in which he wryly affirmed making the comment while president.

Chapter 2, Historical Background to 1972

1. Lewis Hanke, *The Spanish Struggle for Justice in the Conquest of America* (Boston: Little, Brown and Company, 1965), p. 11.
2. Ibid., p. 12.
3. Jorge Juan and Antonio de Ulloa, *A Voyage to South America* (New York: Alfred A. Knopf, 1964), pp. 140–56.
4. Friedrich Hassaurek, *Four Years among the Ecuadorians* (Carbondale: Southern Illinois University Press, 1967), pp. 169–71.
5. Ibid., p. 107.
6. José María Velasco Ibarra, *Obras Completas* (Quito: Ed. Lexigama, 1974), pp. 380–81.
7. *National Geographic* (February 1968): 275–76.

Chapter 3, The Oil Era

1. Author's interview with Galo Plaza Lasso in the spring of 1979.
2. Undated press release titled CONAIE, CONFEDERACIÓN DE NACIONALIDADES INDÍGENAS DEL ECUADOR (Quito). Press releases and CONAIE's newsletters, published sporadically, are available at CONAIE headquarters in Quito [Los Granados 2553 y 6 de Diciembre].
3. Author's interview with Gustavo Noboa on 08-10-00.

Chapter 4, The Emergence of the Indian Movement

1. *CONFENIAE, Unidad, Tierra, Justicia y Libertad* (Quito). The organization's policy statements and newsletters are published sporadically and are available at CONFENIAE headquarters in Quito [Av. 6 de Diciembre 159 y Pazmino Of. 408]. Dozens of the statements and newsletters, particularly those published throughout the 1990s, are an invaluable source for the evolution of various regional groups to form CONAIE.

2. Undated 1994 press release titled *CONAIE, CONFEDERACIÓN DE NACIONALIDADES INDÍGENAS DEL ECUADOR* (Quito). See also the article by Chris Jochnick, director of the Center for Economic and Social Rights based in New York City, in the *Multinational Monitor* (Washington, DC), 02-95, a monthly newsmagazine that tracks the activities of multinational corporations.

3. *El Comercio* (Quito) 01-23-00 and *Reuters*, Alistair Scrutton, Quito, 1-23-00.

4. *El Comercio* (Quito) 01-23-00 and *Diario La Hora* (Quito) 01-23-00.

5. The quote from Leslie Alexander is from the author's interview with Maríana Neira of *Vistazo* (Quito) in August 2000. See also *Militant*, Hilda Cuzco's article, 02-24-97.

6. *New York Times*, Larry Rohter, 01-27-00.

7. The three quotes from Luis Macas came from an undated press release titled *CONAIE, CONFEDERACIÓN DE NACIONALIDADES INDÍGENAS DEL ECUADOR* (Quito).

8. Undated policy statement titled *Que es la FEINE?* (Quito) available at FEINE headquarters in Quito [Isla San Cristobal y Yasuni].

9. *New York Times*, Larry Rohter, 01-27-00.

10. The José María Velasco Ibarra quote by Luis Macas came from an undated press release titled *CONAIE, CONFEDERACIÓN DE NACIONALIDADES INDÍGENAS DEL ECUADOR* (Quito).

11. Undated press release titled *CONAIE, CONFEDERACIÓN DE NACIONALIDADES INDÍGENAS DEL ECUADOR* (Quito).

12. Author's interview with Blanca Chancoso in Quito on 08-07-00.

13. For the reactions of *El Comercio* and others to CONAIE's actions and policy declarations, see *Inter-Press Service*, Quito, 05-30-91.

14. The two quotes from Antonio Vargas and the quote from Nina Pacari came from *Washington Post*, Stephen Buckley, 01-27-00.

15. Enrique Ayala Mora et al., *Pueblos indios, estado y derecho* (Quito: Editora Nacional, 1992), and Luis Maldonado, coord., *Las nacionalidades indígenas en el Ecuador: nuestro proceso organizativo* (Quito: 2d ed., Abya Yala, 1989).

16. *New York Times*, Larry Rohter, 01-25-00.

17. Ibid., 01-27-00.

Chapter 5, Bucaram, Arteaga, and Alarcón

1. Abdalá Bucaram, *Golpe de Estado* (Guayaquil, PREdiciones, 1998), and *Associated Press*, Monte Hayes, Quito, 02-10-97. *El Comercio* (Quito) newspaper prepared in February 1997 and has at its Quito office an excellent 26-page review of Bucaram's tenure titled *El Bucaramato*.

2. Rosalía Arteaga, *La presidenta: el secuestro de una protesta* (Quito: Edino, 1997).

3. Abdalá Bucaram's conservative policies were perceived as "shock therapy." See *New York Transfer News Collective* (The Blythe Systems), 02-18-97.

4. *El Comercio* (Quito) 02-15-97. See also *Militant*, Hilda Cuzco's article, 02-24-97.

5. Bucaram, *Golpe de Estado.*

6. Author's interview with Deputy Víctor Hugo Sicouret Olvera on 08-09-00.

7. Arteaga, *La presidenta.*

8. The quotations from Leslie Alexander are from the author's interviews with Maríana Neira of *Vistazo* in August 2000. See also *Militant*, Hilda Cuzco's article, 02-24-97.

9. The three quotes from Abdalá Bucaram are from his *Golpe de Estado.*

10. Press release of 01-31-95 titled *CONAIE, CONFEDERACIÓN DE NACIONALIDADES INDÍGENAS DEL ECUADOR* (Quito).

11. Bucaram, *Golpe de Estado.*

12. The quotation from León Febres Cordero was referred to again in *Reuters*, 02-05-97.

13. Arteaga, *La presidenta.*

14. The leaflets were printed in *New York Transfer News Collective* (The Blythe Systems), Michael Pearlman, 02-18-97 and in *El Comercio*, 02-05-97. See also *Militant*, Hilda Cuzco's article, 02-24-97.

15. Arteaga, *La presidenta.*

16. The two quotes from Abdalá Bucaram are from *El Comercio* (Quito) 02-02-97 and from his *Golpe de Estado.*

17. Arteaga, *La presidenta.*

18. Ibid.

19. Ibid.

20. Bucaram, *Golpe de Estado.*

21. The congressional motion deposing Bucaram and the quotation from General Paco Moncayo can be found in *El Comercio* (Quito) 02-07-97.

22. Arteaga, *La presidenta.*

23. Ibid.

24. Ibid.

25. Ibid.

26. Arteaga, *La presidenta;* The quotes from Abdalá Bucaram are from *El Telegrafo* (Guayaquil) and *El Universo* (Guayaquil) 02-08-00.

27. Bucaram, *Golpe de Estado.*

28. The quotes from *Diario Hoy* (Quito) are from the edition of 02-08-97.

29. Bucaram, *Golpe de Estado.*

30. The several quotes from Rosalía Arteaga are from her *La presidenta* and *Associated Press*, Monte Hayes, Quito, 02-10-97. A synopsis of events is in *Diario Hoy* (Quito) 09-21-97. *El Comercio's* review of events titled *El Bucaramato* reports that prior to Arteaga being sworn in as president, General Paco Moncayo assured Quito mayor Jamil Mahuad that he would force her to resign within a few days of her taking office.

31. The quotes from Abdalá Bucaram and Alfredo Adum are from *El Universo* (Guayaquil) 02-10-97, and *Associated Press*, Monte Hayes, Quito, 02-10-97.

32. The several quotes from Rosalía Arteaga are from her *La presidenta.*
33. See Bucaram, *Golpe de Estado.*
34. *El Comercio* (Quito) 02-15-97.
35. Ibid., 02-20-97.
36. Ibid., 04-20-97 to 04-30-97.

Chapter 6, Jamil Mahuad

1. *El Comercio* (Quito) 02-06-00.
2. The election figures are from Enrique Ayala Mora, ed., *Nueva Historia del Ecuador,* vol. 11, *Epoca Republicana V* (Quito: Editora Nacional, 1991).
3. The quotes from the various candidates and their policy positions are from literature from their campaign headquarters in Quito; from various issues of *Vistazo* (Quito) from 01- through 05-1998, most notably the issue of 05-21-98; and from numerous issues published during the same time of *El Comercio* (Quito), *El Universo* (Guayaquil), *El Telégrafo* (Guayaquil), *Diario Hoy* (Quito), *Diario La Hora* (Quito), and *El Mercurio* (Cuenca). A summary profile of Mahuad is in *Reuters,* Quito, 01-21-00.
4. Álvaro Noboa's campaign spending is analyzed in *Vistazo* (Quito) 03-02-00.
5. The quote from Rosendo Rojas is from *Reuters,* 07-12-00. General Paco Moncayo agreed. See Moncayo's account in Napoleón Saltos, *La rebelión del arcoiris: testimonios y análisis* (Quito: Fundación José Peralta, 2000), p. 73. The quote from Mahuad is from *El Comercio* (Quito) 06-15-98. The quote from Nebot is from *El Comercio* (Quito) 06-01-98.
6. *El Comercio* (Quito) 07-16-98 and *Associated Press,* José Velázquez, Quito, 07-15-98.
7. Vladimiro Álvarez, *El golpe detrás de los ponchos* (Quito: Edino, 2001).
8. Allen Gerlach, "Peru's Presidential Election and the War with Ecuador," *Contemporary Review* (London), June 1995.
9. Carlos Mendoza, *Quién derrocó a Mahuad?* (Quito: Edi Ecuatorial, 2001).
10. Bucaram, *Golpe de Estado.*
11. For the attitude of Colonel Fausto Cobo and for Jamil Mahuad on military budget cuts in both Ecuador and Peru and the peace treaty, see *Reuters,* Alistair Scrutton, Quito, 01-25-00.
12. The banking problem from Mahuad's perspective is discussed in Álvarez, *El golpe detrás de los ponchos.* The AGD and the banking crisis are discussed with some perspective in *Vistazo* (Quito) 04-06-00, 06-15-00, and 07-20-00.
13. *El Telegrafo* (Guayaquil) 11-15-98.
14. Álvarez, *El golpe detrás de los ponchos* (Quito, Edino, 2001).
15. The two quotes on the banking crisis are from Mendoza, *Quién derrocó a Mahuad?*
16. Álvarez, *El golpe detrás de los ponchos.*
17. *Washington Post,* 04-23-00.
18. The quote from General Carlos Mendoza is from his *Quién derrocó a Mahuad?*; the quote from Vice President Noboa is from *El Universo* (Guayaquil), 04-21-99.
19. Mendoza, *Quién derrocó a Mahuad?*
20. *Diario La Hora* (Quito) 06-29-99.

21. *Reuters*, 07-02-99.
22. *El Telégrafo* (Guayaquil) 07-14-99.
23. Ibid.
24. Mendoza, *Quién derrocó a Mahuad?*
25. *Associated Press*, Carlos Cisternas, Quito, 04-20-99.
26. *El Universo* (Guayaquil) 07-19-99.
27. Mendoza, *Quién derrocó a Mahuad?*
28. The quote from Mahuad is from *Associated Press*, Carlos Cisternas, Quito, 04-20-99.
29. Ecuador and Plan Colombia are treated in Kintto Lucas, *Plan Colombia: la paz armada* (Quito: Planeta, 2000).
30. The quotes from General Carlos Mendoza and Gustavo Noboa are from Mendoza, *Quién derrocó a Mahuad?* See also *Vistazo* (Quito) 10-05 and 19-00.
31. The quote from Rodrigo Borja Cevallos is from *El Universo* (Guayaquil) 10-11-99. See also *Associated Press*, Monte Hayes, Quito, 01-22-00.
32. For Fernando Aspiazu and Jamil Mahuad's refusal to answer Aspiazu's questions before a judge, see *Reuters*, Quito, 11-09-99, and *El Comercio* (Quito) 11-10-99 and 02-27-00.
33. The quote from Borja is from Mendoza, *Quién derrocó a Mahuad?* See also *Reuters*, Quito, 11-09-00.
34. *El Universo* (Guayaquil) 10-24-99.
35. Mendoza, *Quién derrocó a Mahuad?*
36. *El Universo* (Guayaquil) 10-11-99. See also *Associated Press*, Monte Hayes, Quito, 01-22-00.
37. For Mahuad's budget and economic projections, see *Reuters*, Quito, 11-13-99.
38. Undated 1999 press release titled *CONAIE, CONFEDERACIÓN DE NACIONALIDADES INDÍGENAS DEL ECUADOR* (Quito).
39. The quote from Benjamín Ortiz is from *Reuters*, Quito, 11-13-99. Ecuador, the Manta military base, and Plan Colombia are treated in *New York Times*, 12-31-00 and Lucas, *Plan Colombia*.
40. *Expreso* (Quito) 11-15-99.
41. Ibid., 11-22-99 and 12-29-99.
42. Ibid., 11-18-99.
43. The quote from Luis Betun is from *El Comercio* (Quito) 11-22-99, and *Reuters*, Quito, 11-22-99. See also *Reuters*, Quito, 11-21-99.
44. Mendoza, *Quién derrocó a Mahuad?* and *Reuters*, Quito, 11-20-99. See also *Vistazo* (Quito) 02-03-00, and *El Comercio* (Quito) 02-26-00.
45. *El Comercio* (Quito) 12-31-99 and *Reuters*, Quito, 01-21-00.
46. *Reuters*, 01-10-00. See also Borjas's statements in Francisco Herrera Arqúz, *Los golpes del poder . . . al aire: el 21 de enero a través de la radio* (Quito: Abya Yala, 2001), pp. 114–19.
47. *Reuters*, Gustavo Oviedo, Quito, 01-08-00. See also *Vistazo* (Quito) 02-03-00.
48. Mendoza, *Quién derrocó a Mahuad?* See also *Washington Post*, Stephen Buckley, 01-25-00.
49. Colonel Jorge Luis Brito's reflections on Mahuad's fall, with the perspective of almost a year, are in *El Universo* (Guayaquil) 01-19-01. See also *Vistazo* (Quito) 02-03-00.
50. *Reuters*, Gustavo Oviedo, Quito, 01-08-00.
51. Ibid.

52. The quote from Apolinario Quishpe is from *New York Times*, Larry Rohter, 01-27-00. See also *Reuters*, Gustavo Oviedo, Quito, 01-08-00. The quote from Augusto Aguirre is from *Reuters*, Gustavo Oviedo, Quito, 01-11-00 and *New York Times*, Larry Rohter, Quito, 01-11-00.

53. *Reuters*, Gustavo Oviedo, Quito, 01-08 and 09-00, and *Reuters*, Stephen Brown, Buenos Aires, 01-11-00.

54. The quotes from Ramiro Crespo and León Roldós are from *New York Times*, Larry Rohter, 01-16-00. Larry Summers's and John S. Reed's comments are from *Reuters*, Gustavo Oviedo, Quito, 01-08-00, and *Reuters*, Stephen Brown, Buenos Aires, 01-11-00. The quote from Guillermo Chapman is from *Reuters*, Stephen Brown, Buenos Aires, 01-11-00. The quote from Arcesio Vega is from ibid.

55. *Reuters*, Stephen Brown, Buenos Aires, 01-11-00. See also *Reuters*, Alistair Scrutton, Quito, 01-13-00.

56. *Associated Press*, Monte Hayes, Quito, 01-22-00. See also *Ecuador Weekly Report* (Quito) 01-11 and 17-2000. Mahuad's opposition to dollarization in late December is discussed in *El Comercio* (Quito) 02-26-00.

57. Colonel Jorge Luis Brito's reflections on Mahuad's fall, with the perspective of almost a year, are in *El Universo* (Guayaquil) 01-19-01.

58. A summary of General Carlos Mendoza's career is in *Reuters*, Quito, 01-22-00.

59. The quote from CONAIE and the organization's demands are from an undated press release of 01-00, which is strikingly similar to the press release of 03-15-99 that is also titled *CONAIE, CONFEDERACIÓN DE NACIONALIDADES INDÍGENAS DEL ECUADOR* (Quito).

60. See Jorge Brito's account in *El Universo* (Guayaquil) 01-19-01. See also Saltos, *La rebelión del arcoiris*, pp. 187–92.

61. *Vistazo* (Quito) 02-03-00. See also *New York Times*, Larry Rohter, 01-27-00.

62. *Associated Press*, Monte Hayes, Quito, 01-23-00.

63. *Associated Press*, Frank Bajak, Quito, 01-23-00.

64. Bucaram, *Golpe de Estado*.

65. The flight of Ecuadoreans is discussed in *Vistazo* (Quito) 06-01-00.

66. *Associated Press*, Carlos Cisternas, Quito, 04-20-99, and *Reuters*, Quito, 04-26-99. For an elaboration on the dismal analysis, see *Reuters*, Gustavo Oviedo, Quito, 12-09-99.

67. The quote from General Paco Moncayo is from *Reuters*, 12-10-99. Moncayo's views on the role of the military in social and economic development are presented in his *Fuerzas armadas y sociedad* (Quito: Corporación Editora Nacional, 1995).

68. Author's interview with Jim Brown in Chota on 08-12-00.

69. *New York Times*, Larry Rohter, 01-27-00.

70. *Associated Press*, Carlos Cisternas, Quito, 04-20-99.

Chapter 7, Levantamiento Indígena

1. *El Comercio* (Quito) 11-09-00 and *Reuters*, Gustavo Oviedo, Quito, 11-08-00.

2. *Inter-Press Service*, Kintto Lucas, Quito, 02-02-00, *Reuters*, Carlos DeJuana, Quito, 01-15-00, and *Reuters*, Alistair Scrutton, Quito, 01-16-00.

3. *Reuters*, Carlos DeJuana, Quito, 01-15-00.

4. *Inter-Press Service*, Kintto Lucas, Quito, 02-02-00.

5. *El Comercio* (Quito) 01-20-00 and *Reuters*, 01-19-00.

6. Mendoza, *Quién derrocó a Mahuad?*

7. The pamphlets dropped on El Arbolito park are described in *Agencia Informativa Pulsar*, Marlon Carrión C., 01-20-00, and in *Reuters*, Quito, 01-19-00.

8. *El Comercio* (Quito) 02-23 to 28-00.

9. Mendoza, *Quién derrocó a Mahuad?*

10. The exchange between Jamil Mahuad and General Carlos Mendoza is from Mendoza, *Quién derrocó a Mahuad?* and *El Comercio* (Quito) 02-23 to 28-00.

11. Undated press release of 01-00 titled *CONAIE, CONFEDERACIÓN DE NACIONALIDADES INDÍGENAS DEL ECUADOR* (Quito). The title of the release can be translated as "Searching for the Coup Participants among the Oligarchy." See also *New York Times*, Larry Rohter, Quito, 01-10-00.

12. The several quotes from General Carlos Mendoza are from Mendoza, *Quién derrocó a Mahuad?*. See also *Vistazo* (Quito) 02-03-00 and *El Comercio* (Quito) 02-23 to 28-00, and *Inter-Press Service*, Kintto Lucas, Quito, 02-02-00. The perspectives of high-ranking police officials on the events of 01-20-00 are in *El Comercio* (Quito), 05-09-00.

13. *El Comercio* (Quito) 02-23 to 28-00.

14. The plans to rescue employees in Congress and Supreme Court are discussed in *El Comercio* (Quito) 02-24-00.

15. From Brito's account in Saltos, *La rebelión del arcoiris*, pp. 187–92, and the letter written by General José Gallardo and published by *El Comercio* (Quito) in *21 de enero: la voragine que acabó con Mahuad* (Quito: El Comercio, 2000), pp. 247–53.

16. *El Comercio* (Quito) 01-15-00 and *Reuters*, Carlos DeJuana, Quito, 01-15-00.

17. Undated press release titled *CONAIE, CONFEDERACIÓN DE NACIONALIDADES INDÍGENAS DEL ECUADOR* (Quito). See also the 5-page bulletin published by the Instituto Científico de Culturas Indígenas, Quito, in 01-00 titled "Crónica del Levantamiento Indígena y de la Sociedad Civil del Ecuador: La Necesidad de Construir una Verdadera Democracia." *El Comercio* (Quito) printed its detailed account of the events of 12-99 and 01-00, "La historia oculta del golpe: informe especial," in its editions of the last week of 02-00. See also *Vistazo* (Quito) 02-03-00.

18. For the views of Captain César Díaz, see *Vistazo* (Quito) 02-03-00, *El Comercio* (Quito) 02-23-00, and *Reuters*, Lima, 01-21-00.

19. For Colonel Lucio Gutíerrez's meetings with CONAIE directors, see *El Comercio* (Quito) 02-23-00 and 03-01-00.

20. *El Comercio* (Quito) 02-23 to 28-00. See also *Vistazo* (Quito) 02-03-00 and *El Comercio* (Quito) 03-19-00. For Gutíerrez's admiration of President Hugo Chávez of Venezuela, see *Diario La Hora* (Quito) 06-02-00.

21. The several quotes from Colonel Lucio Gutiérrez are from *El Comercio* (Quito) 02-23 to 28-00. See also *Vistazo* (Quito) 02-03-00 and *Reuters*, Quito, 01-22-00.

22. *El Comercio* (Quito) 02-23 and 24-00.

23. *Inter-Press Service*, Kintto Lucas, Quito, 02-02-00. The summary of Vargas is from a publication titled *CONAIE, CONFEDERACIÓN DE NACIONALIDADES INDÍGENAS DEL ECUADOR* (Quito) and subtitled

"Perfil: Antonio Vargas," available at CONAIE headquarters. See also *Reuters*, Quito, 01-22-00.

24. Antonio Vargas's background is in the reference cited in note 23; Salvador Quishpe's background is reported in *Vistazo* (Quito) 03-16-00.

25. *Vistazo* (Quito) 02-03-00. See also *Reuters*, Quito, 01-22-00.

26. The quotes from Carlos Solórzano and Colonel Acosta are from *Associated Press*, Frank Bajak, Quito, 01-22-00. See also *El Comercio* (Quito) 02-27-00.

27. General Paco Moncayo's reflections on Mahuad's fall, with the perspective of almost a year, are in *Diario La Hora* (Quito) 01-21-01.

28. The quote from General Carlos Mendoza is from *El Comercio* (Quito) 02-23 to 28-00. See also *Vistazo* (Quito) 02-03-00.

29. *El Comercio* (Quito) 02-23 to 28-00. See also *Vistazo* (Quito) 02-03-00 and *Reuters*, Carlos DeJuana, Quito, 01-21-00.

30. *El Comercio* (Quito) 02-23 to 28-00. See also *El Comercio* (Quito) 03-20-20.

31. The several quotes from General Carlos Mendoza are from Mendoza, *Quién derrocó a Mahuad?*, *El Comercio* (Quito) 02-26-00, *Associated Press*, Frank Bajak, Quito, 1-24-00, *Associated Press*, Monte Hayes, Quito, 01-21-00, and *Reuters*, Quito, 01-21-00.

32. Álvarez, *El golpe detrás de los ponchos*. See also *Reuters*, Carlos DeJuana, Quito, 01-21-00.

33. *Reuters*, Quito, 01-22-00. See also *Associated Press*, Monte Hayes, Quito, 01-21-00, *Reuters*, Carlos DeJuana, Quito, 01-21-00, and *Reuters*, Alistair Scrutton, Quito, 1-25-00.

34. The several quotes from General Carlos Mendoza are from Mendoza, *Quién derrocó a Mahuad?*

35. *Vistazo* (Quito) 02-03-00 and *El Comercio* (Quito) 01-22-00.

36. An army colonel who asked to remain anonymous explained the events in Guayaquil and the role of León Febres Cordero during interviews with the author in August 2000. See also *El Comercio* (Quito) 02-26-00.

37. *Associated Press*, Frank Bajak, Quito, 01-22-00.

38. Ibid. See also the *Militant*, Hilda Cuzco's article, 02-24-97, 01-22-00.

39. *Quién derrocó a Mahuad?*, *Reuters*, Carlos DeJuana, Quito, 01-21-00, and *Washington Post*, Stephen Buckley, 01-26-00.

40. Author's interview with Antonio Vargas on 08-07-00.

41. For quotations of statements issued by the U.S. government see *Reuters*, Lorraine Orlandi, Mexico City, 01-21-00, *Reuters*, Carlos DeJuana, Quito, 01-21-00, and *Washington Post*, Stephen Buckley, 01-26-00

42. *Associated Press*, Harry Dunphy, Washington, DC, 01-21-00.

43. The quotes from Kofi Annan, John Battle, and the French Foreign Ministry are from *Reuters*, Gilbert Le Gras, 01-22-00.

44. *Associated Press*, Monte Hayes, Quito, 01-24-00. See also *Militant*, Hilda Cuzco's article, 02-24-97, 01-22-00.

45. *New York Times*, Larry Rohter, 01-16-00. The author was told the same by numerous Ecuadorean politicians and journalists in August 2000.

46. For the view of Colonel Fausto Cobo, see *Reuters*, Carlos DeJuana, Quito, 01-21-00. See also *Vistazo* (Quito) 02-03-01.

47. Jamil Mahuad refers to military budget cuts in both Ecuador and Peru and the peace treaty in *Reuters*, Alistair Scrutton, Quito, 01-25-00.

48. *Reuters*, Alistair Scrutton, Quito, 01-24-00.

49. *Reuters*, 06-24-00. See also *El Universo* (Guayaquil) 06-24-00.

50. *Associated Press*, Monte Hayes, Quito, 01-24-00.

51. Undated 2000 press release titled *CONAIE, CONFEDERACIÓN DE NACIONALIDADES INDÍGENAS DEL ECUADOR* (Quito). See also Mendoza, *Quién derrocó a Mahuad?*

52. The several quotes from General Carlos Mendoza are from Mendoza, *Quién derrocó a Mahuad?*

53. Author's interview with Colonel Gutiérrez in Quito in July 2001.

54. Vargas's reflections on Mahaud's fall, with the perspective of a little over a year, are in an undated press release of early January 2001 titled *CONAIE, CONFEDERACIÓN DE NACIONALIDADES INDÍGENAS DEL ECUADOR* (Quito). The release is an interview conducted by Karina Aviles and Miguel Angel Velázquez. See also *Diario La Hora* (Quito) 01-21-01.

55. *Vistazo* (Quito) 05-18-00.

56. *El Comercio* (Quito) 01-22-00.

57. *El Comercio* (Quito), ed., *21 de enero*, p. 75.

58. Ibid., *El Comercio* (Quito) 02-28-00. See also *Reuters*, Quito, 01-22-00.

59. *Reuters*, 01-24-00. See also *New York Times*, Larry Rohter, 01-27-00 and *El Comercio* (Quito) 02-28-00)

60. The quotes from Carlos Solórzano and General Carlos Mendoza are from *El Comercio* (Quito) ed., *21 de enero*, p. 77, and *El Comercio* (Quito) 02-28-00. See also *Associated Press*, Monte Hayes, Quito, 01-21-00.

61. The quotes from Antonio Vargas and Jorge Loor are from *Reuters*, 01-22-00.

62. Mendoz, *Quién derrocó a Mahuad?*. See also *New York Times*, Larry Rohter, Lima, 01-22-00, *Washington Post*, Stephen Buckley, 01-26-00, *Associated Press*, Frank Bajak, Quito, 01-24-00, *Associated Press*, Monte Hayes, Quito, 01-21-00, and *Reuters*, Carlos DeJuana, Quito, 01-22-00.

63. From Brito's account in Saltos, *La rebelión del arcoiris*, pp. 187–92.

64. Author's interview with César Gaviria in Albuquerque, New Mexico, on 02-22-00.

65. *Associated Press*, Frank Bajak, Quito, 01-22-00.

66. *Associated Press*, Monte Hayes, Quito, 01-24-00.

67. *Reuters*, 01-25-00. Mahuad's view is shared by Álvarez in Álvarez, *El golpe detrás de los ponchos*.

68. *El Comercio* (Quito) 02-27-00.

69. *Inter-Press Service*, Kintto Lucas, Quito, 02-02-00 and *Washington Post*, Stephen Buckley, 01-27-00.

70. *Inter-Press Service*, Kintto Lucas, Quito, 02-02-00 and Vargas's account in Saltos, *La rebelión del arcoiris*, pp. 297–302. Vargas's reflections on Mahaud's fall, with the perspective of a little over a year, are in an undated press release of early January 2001 titled *CONAIE, CONFEDERACIÓN DE NACIONALIDADES INDÍGENAS DEL ECUADOR* (Quito). The release is an interview conducted by Karina Aviles and Miguel Angel Velázquez. See also *Diario La Hora* (Quito) 01-21-01.

71. *El Comercio* (Quito) 02-17-00. See also *Ecuador Weekly Report* (Quito) 01-24-00.

Chapter 8, Gustavo Noboa

1. *El Comercio* (Quito), 01-23-00.

2. The several quotes from Gustavo Noboa are from *New York Times*, Larry Rohter, Quito, 01-25-00, *Diario la Hora* (Quito), 06-10-00, *Reuters*, Carlos

DeJuana, Quito, 01-23-00, and *Reuters*, Quito, 01-26-00, and *El Comercio* (Quito) 01-25-00. See also *El Comercio* (Quito) 02-03-00. For the legal and practical considerations involved in Noboa assuming power, see *Ecuador Weekly Report* (Quito) 01-24-00.

3. For the views of Gustavo Noboa see *Vistazo* (Quito) 04-19-00, 05-04-00, 05-18-00, 06-15-00, and 10-05-00. See also *Reuters*, 09-10-00, and *Vistazo* (Quito) 12-14-00.

4. The quotes from Francisco Huerta and the minister of finance are from *El Comercio* (Quito) 02-08-00 and 03-02-00 and *Reuters*, Quito, 02-06-00.

5. *New York Times*, Larry Rohter, Quito, 01-25-00. See also *Reuters*, Washington, DC, 01-22-01, *Reuters*, Alistair Scrutton, Quito, 01-22-00, *Reuters*, Buenos Aires, 01-23-00, *Reuters*, Quito, 01-26-00, and *El Comercio* (Quito) 02-16-00.

6. See *Reuters*, Alejandro Aguirre, Quito, 04-09-00. See also Gutíerrez's statements in Herrera Arduz, *Los golpes del poder . . . al aire*, p. 96–97.

7. *El Comercio* (Quito) 03-30-00. See also *Associated Press*, Quito, 03-29-00.

8. *El Comercio* (Quito) 04-13-00 and 05-09-00. See also *El Comercio* (Quito) 02-04-00 and *Reuters*, Quito, 04-12-00.

9. *El Universo* (Guayaquil) 06-02-00.

10. *Associated Press*, Monte Hayes, Quito, 01-24-00.

11. *El Comercio* (Quito) is from the 01-23-00 edition.

12. *Reuters*, Alistair Scrutton, Quito, 01-22-00 and 01-23-00.

13. The several quotes from Antonio Vargas are from *Reuters*, Alistair Scrutton, Quito, 01-23-00, *Inter-Press Service*, Kintto Lucas, Quito, 02-02-00, *El Comercio* (Quito) 01-23-00, 02-20-00, and 05-06-00, *Associated Press*, Monte Hayes, Quito, 01-24-00 and 01-29-00, *The Associated Press*, Quito, 01-26-00, and *Washington Post*, Stephen Buckley, 01-26-00.

14. *Vistazo* (Quito) 02-03-00.

15. For Antonio Vargas's plebiscite proposal see *Ecuador Weekly Report* (Quito) 02-14-00 and *Inter-Press Service*, Kintto Lucas, Quito, 02-02-00. See also *Vistazo* (Quito) 03-16-00, *El Comercio* (Quito) 02-13, 18-00, and 03-05-00, *Reuters*, Virginia Burgos, Guayaquil, 02-06-00, *Reuters*, María José González, Lima, 02-04-00, *Reuters*, Quito, 02-09-00, and *Reuters*, Alejandro Aguirre, Quito, 08-02-00.

16. The two quotes from César Gaviria are from *El Comercio* (Quito) 02-16-00 and 02-17-00. See also *Reuters*, 02-16-00.

17. *Reuters*, María José González, Lima, 02-04-00.

18. *Reuters*, Alejandro Aguirre, Quito, 05-02-00.

19. Ley Trole I is summarized in *Reuters*, Quito, 03-01-00. See also *El Comercio* (Quito) 03-07 through 09-00. The quote from Stanley Fischer is from *Reuters*, Quito, 03-17-00.

20. For the views of Edwin Truman, Stanley Fischer, and other IMF officials (Thomas Pickering and others), see *El Comercio* (Quito) 02-16 to l8-00, *Reuters*, Quito, 02-16-00, *Reuters* (Quito), Gustavo Oviedo, 03-29-00, and *Associated Press*, Harry Dunphy, Washington, DC, 03-09-00. See also *Ecuador Weekly Report* (Quito) 02-28-00.

21. The quotes from Heinz Moeller and Gladys Silva are from *Washington Post*, Anthony Faiola, Quito, 04-09-00.

22. The opposition of the Indian and Social Movements to the dollar plan is summarized in *Reuters*, 03-21-00.

23. *Reuters*, 03-25-00.

24. *El Universo* (Guayaquil) 05-06-00.

25. For an analysis of the relationship between increased oil prices and government revenue (the government received $80 million more per year for every 1-dollar increase in the price of a barrel of crude oil), see *El Comercio* (Quito) 02-19-01.

26. *El Comercio* (Quito) 04-27 to 30-00.

27. Ibid. See also *El Comercio* (Quito) 04-27-00 and *Reuters*, Quito, 05-10-00.

28. The quote from Salvador Quishpe is from *El Comercio* (Quito) 04-27-00. See also the edition of 02-20-00.

29. The questions were raised by *El Comercio* (Quito) in the 02-20-00 edition.

30. *El Universo* (Guayaquil) 05-03-00 and 05-04-00.

31. *El Universo* (Guayaquil) 05-07-00.

32. The several quotes from Gustavo Noboa are from *El Comercio* (Quito) 05-12-00. See also *Reuters*, 05-12-00.

33. *El Universo* (Guayaquil) 05-31-00.

34. *Vistazo* (Quito) 06-15-00. See also the 04-19-00 edition. See also *El Comercio* (Quito) 08-03-00 and *Reuters*, Alejandro Aguirre, Quito, 08-02-00.

35. *Reuters*, Alejandro Aguirre, Quito, 08-02-00. See also *Reuters*, 06-11-00, *Diario La Hora* (Quito) 06-25-00, and *El Comercio* (Quito) 08-03-00.

36. *El Telégrafo* (Guayaquil) 05-01-00.

37. *Diario La Hora* (Quito) 05-03-00. For the demands of the teachers, see *Reuters*, Caracas, 04-27-00, and *Reuters*, Mario Naranjo, Quito, 05-17-00.

38. *El Comercio* (Quito) 06-22 and 23-00 and *Reuters*, 06-22-01.

39. *Reuters*, 04-20-00.

40. *El Comercio* (Quito) 05-08 and 11-00.

41. Ibid., 05-17-00. The quote from Jorge Guzmán is from Ibid.

41.1. *Reuters*, Quito, 05-25-00.

42. *El Comercio* (Quito) 05-12-00.

43. The quotes from Wilson Álvarez, Napoleón Saltos, and Salvador Quishpe are from *El Comercio*, 05-13 through 17-00.

44. *Reuters*, 04-20-00. See also *Diario La Hora* (Quito), 07-03-00.

45. The quotes from Telmo Sandoval, Carlos Calle, José Gallardo, Jorge Villarroel, Carlos Mendoza, Antonio Vargas, and Lucio Gutiérrez are from *El Telégrafo* (Guayaquil), *El Comercio* (Quito), *El Universo* (Guayaquil), and *Diario La Hora* (Quito) 05-04-00 and 05-05-00.

46. *Diario La Hora* (Quito) 06-02-00 and *El Telégrafo* (Guayaquil) 05-05-00 and 06-02-00. See also *El Comercio* (Quito) 05-05-00 and *Vistazo* (Quito) 07-06-00.

47. *El Comercio* (Quito) 05-12-00.

48. *El Universo* (Guayaquil) 05-08-00.

49. The quote from Jaime Nebot is from *El Comercio* (Quito) 05-09-00. See *Diario La Hora* (Quito) 06-01-00 for the Social Christian Party's stance on amnesty.

50. The quotes from Abdalá Bucaram and León Febres Cordero are from *El Universo* (Guayaquil) 06-02-00 and *Reuters*, 05-31-00. See also *El Comercio* (Quito) 05-16-00 and *El Telégrafo* (Guayaquil) 04-14-00.

51. *Reuters*, 05-31-00 and *El Comercio* (Quito) 06-01-00.

52. *El Telégrafo* (Guayaquil) 06-02-00.

53. *El Telégrafo* (Guayaquil) 06-02-00, *El Universo* (Guayaquil) and *El Comercio* (Quito) of 05-02-00 and 06-01, 02 and 04-00, *Diario La Hora* (Quito) 06-02-00, *Christian Science Monitor*, Howard LaFranchi, Caracas, 06-01-00.

54. *El Universo* (Guayaquil) 06-02-00. See also *El Universo* (Guayaquil) 05-30 and 31-00 for comments by Abdalá Bucaram and his son Jacobo, and Bucaram, *Golpe de Estado*.

55. *El Universo* (Guayaquil) 06-22-00, *Diario La Hora* (Quito) 06-02-00, *El Telegrafo* (Guayaquil) 06-02-00.

56. See in particular *El Comercio* (Quito) 02-26-00, *El Comercio* (Quito) 07-19-00, *Associated Press*, Quito, 02-10-00, *El Mercurio* (Cuenca) 05-16 and 17-00, and *Ecuador Weekly Report* (Quito) 07-24-00. Numerous comments about Plan Colombia's impact on Ecuador are reported in *Vistazo* (Quito) 04-19-00, *Ecuador Weekly Report* (Quito) 03-13-00, *Reuters*, Pablo Garibian, 04-19-00, *Diario La Hora* (Quito) 07-19-00, and *El Universo* (Guayaquil) 07-08-00.

57. The quotes from Bishop Gonzalo López, Luis Villacis, and Gustavo Noboa are from *Miami Herald*, Juan O. Tamayo, 11-12-00. See also *Reuters*, 07-10-00, *El Mercurio* (Cuenca) 05-16 and 17-00, and *New York Times*, Larry Rohter, 01-08-01.

58. An extensive analysis of the election results appears in *El Comercio* (Quito) 05-22 and 23-00.

59. *El Telegrafo* (Guayaquil) 07-14-00.

60. Author's interview with González on 08-09-00.

61. Author's interview with Sicouret on 08-09-00. See also *Vistazo* (Quito) 08-17-00.

62. *Reuters*, Quito, 09-04-00, and *El Telegrafo* (Guayaquil) 09-04-00. See also *Diario La Hora* (Quito) 08-29-00 and 09-03-00 and *El Telégrafo* (Guayaquil) 08-22 through 24-00.

63. *El Comercio* (Quito) 09-07-00. See also *The Associated Press*, Paisley Dodds, New York, 09-06-00, *El Comercio* (Quito) 09-06-00 and 09-07-00, and *Diario La Hora* (Quito) 09-03-00.

Conclusion

1. *El Comercio* (Quito) 02-02-00.

2. *Vistazo* (Quito) 02-03-00. For other comments from Mahuad, see *Associated Press*, Monte Hayes, Quito, 01-21-00, and *Reuters*, Alistair Scrutton, Quito, 01-25-00.

3. The several quotes from Vladimiro Álvarez Grau are from his *El golpe detrás de los ponchos*.

4. General José Gallardo's analysis is in Saltos, *La rebelión del arcoiris*, p. 105.

5. Mendoza, *Quién derrocó a Mahuad?*

6. See Moncayo's account in Saltos, *La rebelión del arcoiris*, pp. 71–111, and in Herrera Arduz, *Los golpes del poder . . . al aire*, p. 113.

7. *Inter-Press Service*, Kintto Lucas, Quito, 02-02-00. See Vargas's account in Saltos, *La rebelión del arcoiris*, pp. 297–302. Blanca Chancosa of CONAIE ageed with Vargas's analysis in an interview with the author on 08-07-00. Thoughts of various indigenous leaders, including Antonio Vargas, Blanca Chancosa, Salvador Quishpe, Miguel Lluco, and Napoleón Saltos of the CSM appear in *El Comercio* (Quito) 02-20-00.

8. Author's interview with Colonel Lucio Gutiérrez on 07-07-01. See Gutíerrez's account in Saltos, *La rebelión del arcoiris*, pp. 287–95.

9. *Reuters*, Quito, 01-21-00.

10. *New York Times*, Larry Rohter, Quito, 01-25-00 and *Diario La Hora* (Quito), 06-10-00,

11. From Moncayo's account in Saltos, *La rebelión del arcoiris*, pp. 71–111.

12. Álvarez, *Quién derrocó a Mahuad?*

13. From Vargas's account in Saltos, *La rebelión del arcoiris*, pp. 297–302.

Bibliography

Much of what I say is based on the research of others, including historians, social scientists from various disciplines, journalists, and other writers. For the February 1997 and the January 2000 movements, each of which toppled a president, major sources include the Ecuadorean newspapers *El Comercio* (Quito), *El Universo* (Guayaquil), *El Telégrafo* (Guayaquil), *Expreso* (Guayaquil), *Diario Hoy* (Quito), *Diario La Hora* (Quito), *El Mercurio* (Cuenca), and the twice-monthly magazine *Vistazo* (Quito and Guayaquil) as well as *Ecuador Weekly Report* (Quito). The *New York Times* and *Washington Post* were consulted along with Associated Press and Reuters news stories.

The perspectives of CONAIE, the Confederation of Indigenous Nationalities of Ecuador, were gleaned from member interviews, press releases, and the organization's various publications, including its monthly newsletter, *Nacionalidades Indias*. Interviews of active politicians and U.S. embassy briefings in Quito were useful largely for basic overviews as opposed to providing detailed information on events.

Election figures are from the Tribunal Electoral in Quito as well as from Enrique Ayala Mora, editor, *Nueva Historia del Ecuador*, vol. 11, *Epoca Republicana V* (Quito: Editora Nacional, 1991).

Most of the statistics on petroleum production and prices were provided by OLADE, the Organización Latinoamericana de Energía. The numbers for inflation, economic growth, and government income and expenditures are from a variety of publications of the Banco Central del Ecuador, notably its annual publication of statistical data, *Boletín anuario* (Quito), and also from Patricio Almeida Guzmán and Rebeca Almeida Arroba, *Estadísticas Económicas Historicas, 1948–1983* (Quito: 1988), and Alberto Acosta, *Breve historia económica del Ecuador* (Quito: Editora Nacional, 2000). Another annual statistical publication is Instituto Nacional de Estadística del Ecuador, *Anuario de estadística* (Quito). Also consulted was FLASCO, *Los Andes en Cifras* (Quito: 1994).

The basic secondary source for Ecuadorean history is the series of essays in the fifteen volumes of Enrique Ayala Mora, editor, *Nueva Historia del Ecuador* (Quito: Editora Nacional, 1991). Volume 11 covers the most recent period, from 1960 through 1990, stopping about where this work begins its more detailed focus. In addition to the sources listed in those volumes, the list that follows will give the interested reader further material for in-depth looks at the complex forces that have created Ecuador.

Acosta, Alberto. *Breve historia económica del Ecuador* (Quito: Editora Nacional, 2000).
Acosta, Alberto E., et al. *Ecuador: Petróleo y Crisis Económica* (Quito: ILDIS, 1986).
Adoum, Jorge Enrique. *Ecuador: senas particulares* (Quito: Eskeletra Editorial, 2000).
Alban Gómez, Ernesto, et al. *Los indios y el estado-país: pluriculturalidad y multietnicidad en el Ecuador: contribuciones al debate* (Quito: Abya Yala, 1993).
Albornoz Peralta, Oswaldo. *Las Luchas Indígenas en el Ecuador* (Guayaquil: Editorial Claridad, 1976).
______. *Breve síntesis: historia del movimiento obrero ecuatoriano* (Quito: Letra Nueva, 1983).
Alchon, Suzanne Austin. *Native Society and Disease in Colonial Ecuador* (Cambridge: Cambridge University Press, 1991).
Almeida Guzmán, Patricio, and Rebeca Almeida Arroba. *Estadísticas Económicas Historicas, 1948–1983* (Quito: Banco Central del Ecuador, 1988).
Almeida Vinneza, José, et al. *Sismo étnico en el Ecuador* (Quito: Abya Yala, 1993).
______. *Identidades indias en el Ecuador contemporaneo* (Quito: Abya Yala, 1995).
Álvarez Grau, Vladimiro. *El golpe detrás de los ponchos* (Quito: Edino, 2001).
Andrien, Kenneth J. *Andean Worlds: Indigenous History, Culture, and Consciousness under Spanish Rule, 1532–1825* (Albuquerque: University of New Mexico Press, 2001).
______. *The Kingdom of Quito, 1690–1830: The State and Regional Development* (Cambridge: Cambridge University Press, 1995).
Arteaga Serrano, Rosalía. *La presidenta: el secuestro de una protesta* (Quito: Edino, 1997).
Ayala Mora, Enrique. *Ecuador-Perú: historia del conflicto y de la paz* (Quito: Planeta, 1999).
______. *Lucha Política y Origen de los Partidos en el Ecuador* (Quito: Publitecnica, 1978).
______, ed. *Nueva Historia del Ecuador*, 15 vols. (Quito: Editora Nacional, 1991).

______. *Historia de la revolución liberal ecuatoriana* (Quito: Editora Nacional, 1994).

______, et al. *Pueblos indios, estado y derecho* (Quito: Editora Nacional, 1992).

______. *Resumen de Historia del Ecuador* (Quito: Editoria Nacional, 1993).

Baez, Rene, et al. *Ecuador: Pasado y Presente* (Quito: Editorial Ecuador F.B.T., 1975).

Banco Central del Ecuador. *Boletín anuario* (Quito: Banco Central del Ecuador).

Barsky, Osvaldo. *La Reforma Agraria Ecuatoriana* (Quito: Editora Nacional, 1984).

Benitez, Lilyan, and Alicia Garcés. *Culturas Ecuatorianas ayer y hoy* (Quito: Abya Yala, 1998).

Blanksten, George. *Ecuador: Constitutions and Caudillos* (Berkeley: University of California Press, 1951).

Borja N., Raúl. *Comunicación social y pueblos indígenas del Ecuador* (Quito: Abya Yala, 1998).

Botero, Luis Fernando. *Movilización indígena, etnicidad y proceso de simbolización en Ecuador: El caso del lider indígena Lázaro Condo* (Quito: Abya Yala, 2001).

Bucaram Ortíz, Abdalá. *Golpe de Estado* (Guayaquil: PREdiciones, 1998).

Canizares Proaño, Francisco. *La verdadera historia del Ecuador: genocidio en nombre de dios* (Quito: Casa de la Cultura Ecuatoriana, 2000).

Cárdenas Reyes, María Cristina. *Velasco Ibarra: ideología, poder y democracia* (Quito: Editora Nacional, 1991)

Checa Cobo, Marco A. *El Regimen de la Propiedad de la Tierra en El Ecuador* (Quito: Lexigrama, 1973).

Clark, A. Kim. *The Redemptive Work: Railway and Nation in Ecuador, 1895–1930* (Wilmington, DE: Scholarly Resources, 1998).

Confederación de Nacionalidades Indígenas del Ecuador (CONAIE). *Las Nacionalidades Indígenas en el Ecuador: Nuestro Proceso Organizativo* (Quito: CONAIE, 1989).

Cook, Noble David. *Born to Die: Disease and New World Conquest, 1492–1650* (Cambridge: Cambridge University Press, 1998).

Corkill, David. *Ecuador: Fragile Democracy* (London: Latin American Bureau, 1988).

Costales Samaniego, Piedad and Alfred. *Historia Social del Ecuador*, 3 vols. (Quito: IEAG, 1964).

Cueva, Agustín. *El Proceso de Dominación Política en el Ecuador* (Quito: Ed. Critica, 1972).

______. *The Process of Political Domination in Ecuador* (New Brunswick, NJ: Transaction Books, 1982).

Cushner, Nicholas P. *Farm and Factory: The Jesuits and the Development of Agrarian Capitalism in Colonial Quito, 1600–1767* (Albany: State University of New York Press, 1982).

Damerval, Jaime. *Monopolio político del Ecuador* (Quito: Espol, 2000).

Díaz Polanco, Hector. *Indigenous Peoples in Latin America: The Quest for Self-Determination* (Boulder: Westview Press, 1997).
Dietrich, Heinz. *La cuarta via al poder: El 21 de enero desde una perspectiva latinoamericana* (Quito: Abya Yala, 2001).
El Comercio, ed. *21 de enero: la voragine que acabó con Mahuad* (Quito: El Comercio, 2000).
Endara Tomaselli, Lourdes. *La marciano de la esquina: imagen del indio en la prensa ecuatoriana durante el levantamiento de 1990* (Quito: Abya Yala, 1998).
Espinosa Apolo, Manuel. *Los mestizos ecuatorianos y las señas de identidad cultural* (Quito: Taller de Estudios Andinos, 1995).
Ewin, Alexander, ed. *Voice of Indigenous Peoples: Native People Address the United Nations* (Santa Fe: Clear Light Publishers, 1994).
Fitch, John Samuel. *The Military Coup d'Etat as Political Process: Ecuador, 1948–1966* (Baltimore: Johns Hopkins University Press, 1977).
FLASCO. *Los Andes en Cifras* (Quito: FLASCO, 1994).
Galarza, Jaime. *El festín del petróleo* (Quito: Cicetronic, 1972).
Goffin, A. M. *The Rise of Protestant Evangelism in Ecuador, 1895–1990* (Gainesville: University of Florida Press, 1994).
Grijalva Jiménez, Agustín, ed. *Datos básicos de la realidad nacional* (Quito: Editora Nacional, 1998).
Guerra Cáceres, Alejandro. *Esclavos manumitidos durante el govierno del gral. José María Urbina* (Guayaquil: Banco Central del Ecuador, 1997).
Guerrero, Andres. *Haciendas, capital y lucha de clases andinas: disolución de la hacienda serrana y lucha política en los años 1960–64* (Quito: El Conejo, 1983).
Guzmán, Marco Antonio. *Ecuador: La hora trágica, los diferentes rostros de la crisis actual* (Quito: Editora Nacional, 2000).
Hanke, Lewis. *The Spanish Struggle for Justice in the Conquest of America* (Boston: Little, Brown and Company, 1965).
Harner, Michael J. *The Jivaro: People of the Sacred Waterfalls* (Garden City, NY: Anchor Books, 1973).
Hassaurek, Friedrich. *Four Years among the Ecuadorians* (Carbondale: Southern Illinois University Press, 1967).
Hemming, John. *The Conquest of the Incas* (New York: Harcourt Brace & Company, 1970).
Hernández Peñaherrera, Luis, *La guerra del Cenepa: diario de un comandante* (Quito: Editora Nacional, 1998).
Herrera Aráuz, Francisco. *Los golpes del poder . . . al aire: el 21 de enero a través de la radio* (Quito: Abya Yala, 2001).
Horna, Hernán. *La Indianidad: The Indigenous World before Latin Americans* (Princeton: Markus Wiener, 2001).
Hurtado Larrea, Osvaldo. *Dos mundos superpuestos: ensayo de diagnostico de la realidad ecuatoriana* (Quito: INEDES, 1969).
______. *Political Power in Ecuador* (Albuquerque: University of New Mexico Press, 1985).

Ibarra Illanez, Alicia. *Los Indígenas y el Estado en el Ecuador* (Quito: Abya Yala, 1987).

Icaza, Jorge. *The Villagers (Huasipungo)*, translation from the Spanish by Bernard Dulsey of Icaza's 1934 novel (Carbondale: Southern Illinois University Press, 1964).

Instituto Nacional de Estadística del Ecuador. *Anuario de estadística* (Quito: Instituto Nacional de Estadistica).

Isaacs, Anita. *Military Rule and Transition in Ecuador, 1972–92* (Pittsburgh: University of Pittsburgh Press, 1993).

Jaramillo Alvarado, Pío. *El Indio Ecuatoriano*, 2 vols. (Quito: Editora Nacional, 1997).

Kane, Joe. *Savages* (New York: Random House, 1996).

Kimerling, Judith. *Amazon Crude* (New York: Natural Resources Defense Council, 1991).

Larrea Holguin, Juan. *145 Años de Legislación Ecuatoriana, 1830–1975*, 2 vols. (Quito: Banco Central, 1977).

Lascano Palacios, Mario. *21 de Enero: La noche de los coroneles, rebelión de los mandos medios* (Quito: Kess, 2001).

León, Jorge T. *Nuestras Comunidades Ayer y Hoy: Historia de las Comunidades Indígenas de Otavalo, Nucanchic Aillu Llactacuna Naupa, Cunan Pachapash* (Quito: Abya Yala, 1994).

Link, Lilo. *Ecuador: Country of Contrasts* (London: Oxford University Press, 1960).

Lucas, Kintto. *Plan Colombia: la paz armada* (Quito: Planeta, 2000).

Luzuriaga, Carlos, and Clarence Zuvekas, Jr. *Income Distribution and Poverty in Rural Ecuador, 1950–1979* (Tempe: University of Arizona Press, 1980).

Maldonado, Luis, coord. *Las nacionalidades indígenas en el Ecuador: nuestro proceso organizativo* (Quito: 2d ed., Abya Yala, 1989).

Mannheim, Bruce. *The Language of the Inka since the European Invasion* (Austin: University of Texas Press, 1991).

Martz, John D. *Politics and Petroleum in Ecuador* (New Brunswick, NJ: Transaction Books, 1987).

______. *Ecuador: Conflicting Political Culture and the Quest for Progress* (Boston: Allyn and Bacon, 1972).

Mena Villamar, Claudio. *Ecuador a Comienzos de Siglo* (Quito: Abya Yala, 1995).

______. *El Quito rebelde (1809–1812)* (Quito: Abya Yala, 1997).

Mendoza Poveda, Carlos. *Quién derrocó a Mahuad?* (Quito: Edi Ecuatorial, 2001).

Milk, Richard. *Movimiento obrero ecuatoriano: el desafio de la integración* (Quito: Abya Yala, 1997).

Mills, Nick D., Jr. *Crisis, Conflicto y Consenso: Ecuador, 1979–1984* (Quito: Cordes, 1984).

Minchom, Martin. *The People of Quito, 1690–1810: Change and Unrest in the Underclass* (Boulder: Westview Press, 1994).

Moncayo Gallegos, Paco. *Fuerzas armadas y sociedad* (Quito: Editora Nacional, 1995).
Moreno Yañez, Segundo. *Sublevaciones Indígenas en la Audiencia de Quito* (Quito: Universidad Católica, 1978).
Morner, Magnus. *The Andean Past: Land, Societies, Conflicts* (New York: Columbia University Press, 1985).
Morris, Craig, and Adriana von Hagen. *The Inka Empire and Its Andean Origins* (New York: Abbeville Press, 1993).
Múñoz J., Francisco. *Descentralización* (Quito: Tramasocial, 1999).
Murgueytio, Reinaldo. *Tierra, Cultura y Libertad* (Quito: Minerva, 1961).
Needler, Martin C. *Anatomy of a Coup d'Etat: Ecuador, 1963* (Washington, DC: Institute for the Comparative Study of Political Systems, 1964).
Newson, Linda A. *Life and Death in Early Colonial Ecuador* (Norman: University of Oklahoma Press, 1995).
Oberem, Udo. *Conciertos y Huasipungueros en Ecuador* (Quito: Universidad Central, 1967)
Ojeda Segovia, Lautaro. *Encrucijadas y perspectivas de la descentralización en el Ecuador* (Quito: Abya Yala, 1998).
OLADE, Organización Latinoamericana de Energía. OLADE's petroleum figures are available on the Internet at OLADE.com.
Ona Villarreal, Humberto. *Fechas históricas y hombres notables del Ecuador y del mundo* (Quito: Multigráficas, 1979).
Ortíz Villacis, Marcelo. *El control del poder: Ecuador, 1966–1984* (Quito: San Pablo, 1985).
Pareja Diezcanseco, Alfredo. *Breve Historia del Ecuador*, 2 vols. (Quito: Libresa, 1990).
Pérez, Aquiles R. *Las mitas en la real audiencia de Quito* (Guayaquil: Universidad de Guayaquil, 1987).
Phelan, John Liddy. *The Kingdom of Quito in the Seventeenth Century* (Madison: University of Wisconsin Press, 1967).
Philip, George. *Oil and Politics in Latin America: Nationalist Movements and State Companies* (Cambridge: Cambridge University Press, 1982).
Pineo, R. F. *Social and Economic Reform in Ecuador: Life and Work in Guayaquil* (Gainesville: University of Florida Press, 1996).
Plaza, Galo. *Problems of Democracy in Latin America* (Chapel Hill: University of North Carolina Press, 1957).
Powers, Karen Vieira. *Andean Journeys: Migration, Ethnogensis, and the State in Colonial Quito* (Albuquerque: University of New Mexico Press, 1995).
Quintero López, Rafael. *El Mito del Populismo: Análisis de los Fundamentos del Estado Ecuatoriano Moderno (1895–1934)* (Quito: Universidad Central del Ecuador, 1983).
Quintero López, Rafael, and Erika Silva. *Ecuador: una nación en ciernes*, 2 vols. (Quito: Abya Yala, 1991).

Radcliffe, Sarah, and Sallie Westwood. *Remaking the Nation: Place, Identity, and Politics in Latin America* (London and New York: Routledge, 1996).

Redclift, Michael. *Agrarian Reform and Peasant Organizations on the Ecuadorean Coast* (London: Athlone Press, 1978).

Rodas Chávez, Germán. *La izquierda ecuatoriana en el siglo XX* (Quito: Abya Yala, 2000).

Rodríguez, Linda Alexandra. *The Search for Public Policy: Regional Politics and Government Financing in Ecuador, 1830–1940* (Berkeley: University of California Press, 1985).

Rubio Orbe, Gonzalo. *Los indios ecuatorianos: evolución histórica y políticas indígenistas* (Quito: Editora Nacional, 1987).

Sacoto, Antonio. *La novela ecuatoriana, 1970–2000* (Quito: SINAB, 2000).

Salomon, Frank. *Native Lords of Quito in the Age of the Incas: The Political Economy of the North Andean Chiefdoms* (Cambridge: Cambridge University Press, 1986).

Saltos Galarza, Napoleón, ed. *La rebelión del arcoiris: testimonios y análisis* (Quito: Fundación José Peralta, 2000).

Salvador Lara, Jorge. *Historia Contemporánea del Ecuador* (Mexico City: Fondo de la Cultura Mexicana, 1994).

Sampedro V., Francisco. *Geografía Histórica Territorial del Ecuador a 1994* (Quito: DIMAXI, 1994).

Schodt, David W. *Ecuador: An Andean Enigma* (Boulder: Westview Press, 1987).

Sierra Castro, Enrique. *Ecuador, Ecuador: Tu Petróleo, Tu Gente* (Quito: Editorial Cultura y Didactica, 1995).

Teran, Francisco. *Geografía del Ecuador* (Quito: Editora Nacional, 1999).

Tobar Donoso, Julio. *La legislación liberal y la iglesia católica en el Ecuador* (Quito: Producción Gráfica, 2001).

Trabucco, Federico E. *Síntesis Histórica de la República del Ecuador* (Quito: Ed. Santo Domingo, 1968).

Ulloa, Jorge Juan and Antonio de. *A Voyage to South America* (New York: Alfred A. Knopf, 1964).

Urban, Greg, and Joel Sherzer, eds. *Nation-States and Indians in Latin America* (Austin: University of Texas Press, 1991).

Valencia Sala, Gladys. *El mayorazgo en la Audiencia de Quito* (Quito: Abya Yala, 1994).

Van Aken, Mark J. *King of the Night: Juan José Flores and Ecuador, 1824–1864* (Berkeley: University of California Press, 1989).

Velasco, Fernando. *Reforma Agraria y Movimiento Campesino Indígena de la Sierra* (Quito: Editorial El Conejo, 1983).

Velasco Ibarra, José María. *Obras Completas* (Quito: Ed. Lexigama, 1974).

Verdesoto Custode, Luis. *El control social de la gestion pública: linamientos de una política de participación social* (Quito: Abya Yala, 2000).

Villalobos, Fabio. *La industrialización Ecuatoriana: 1976–1983* (Quito: FLASCO-CIPAD, 1987).

Villamizar Herrera, Darío. *Ecuador, 1960–1990: insurgencia democracia y dictadura* (Quito: El Conejo, 1994).
Whitten, Norman E., Jr. *Amazonian Ecuador: An Ethnic Interface in Ecological, Social, and Ideological Perspectives* (Copenhagen: IWGIA, 1978).
______. *Black Frontiersmen: Afro-Hispanic Culture of Ecuador and Colombia* (Prospect Heights: Waveland Press, 1994).
Ycaza, Patricio. *Historia laboral: crónica y debate* (Quito: Editora Nacional, 1995).
Zambrano Ojeda, Noe. *Un pueblo se ha puesto de pie* (Guayaquil: 2000).
Zook, David H., Jr. *Zarumilla-Marañon: The Ecuador-Peru Dispute* (New York: Bookman Associates, 1964).

Index

Abad, Bernardo, 170
Abdum Ziade, Alfredo, 97, 98, 113
Achuar, 4, 8, 10–11, 73, 126. *See also* Indians
Acosta, Patricio, 175, 180
Adum, Alfredo, 83, 102, 105–6
Afro-Ecuadoreans, 7, 13, 27, 28, 71, 77, 159
AGD (Deposit Guarantee Agency), 130–32, 134, 207, 230
Agrarian reform, 31, 63–67, 68, 78, 217; and CONAIE, 65–66, 69, 74, 75, 76, 77, 240, 242
Agriculture, 21–22, 72; bananas, 2, 12, 30, 35, 160; cacao, 22, 23, 24, 26, 27, 35; among Indians, 3, 11, 53, 63–67, 70; rubber, 51
Aguirre, Augusto, 152
Alarcón, Fabián: austerity measures, 111; and corruption, 179, 241–42; foreign debt policies, 108, 110–11; as interim president, 98, 99–100, 101–3, 104–5, 106–7, 108–10, 123–24, 196, 244; national referendum regarding presidency of, 109–10; relationship with Arteaga, 97, 102, 104, 109–10; relationship with Bucaram, 95, 97, 101–2, 108; relationship with Moncayo, 99, 102–3; relations with Peru under, 113; tax policies, 111
Alarcón Falconi, Ruperto, 108
Alava, Milton, 58
Alexander, Leslie, 62, 90, 96, 99, 100, 102–3, 107, 190–91
Alfaro, Eloy, 29
Allende, Salvador, 174
Alliance for Progress, 63
Almagro, Diego de, 18
Alvarado, Pedro de, 18
Alvarado, Rosa, 72
Álvarez, Vladimiro, 123, 132, 133, 236, 240; during Levantamiento Indígena, 165, 167, 168, 170, 184, 185, 186
Álvarez, Wilson, 223
Amazon. *See* Oriente/Amazon
Ampudía, Jarrín, 34
Andean Development Corporation, 43, 130, 140, 215
Anglo Ecuadorean Oil Fields Ltd., 36
Annan, Kofi, 189–90
Argentina, 93, 129, 140, 153; trade bloc, 189
Arízaga, Alfredo, 151, 225, 226
Armed Forces, 82, 123, 242–43; attitudes toward peace treaty with Peru, 127–28, 180, 191; budget of, 26, 90, 92, 127, 174, 180, 191, 192, 222, 246; Indians in the, 191; during Levantamiento Indígena, 155–56, 163, 165, 167–77, 179–80, 181, 183–85, 186, 188, 192–93, 209–10, 227–28, 238–39, 243; and oil revenue, 36; opposed to corruption, 173, 175, 179–80, 181, 191, 192, 227, 237; opposed to privatization, 88; and the poor, 169, 190, 191–92, 240; as reformist, 31, 35–38, 169, 190, 191–93, 240; regional military commanders during Levantamiento Indígena, 184, 188, 197, 200, 202, 236; relationship with Bucaram, 90, 92, 95, 96, 98–101, 102, 105, 107–8, 174, 185, 210, 232, 243; relationship with CONAIE, 72, 155–57, 165,

Armed Forces (*continued*) 166, 167, 168–77, 186–87, 192–94, 199, 201–3, 235; relationship with Mahuad, 127–28, 132–33, 137, 138, 139–40, 142, 146–49, 151, 154, 163, 165, 166–71, 173–75, 184–85, 186, 191–92, 199–200, 202, 208, 210, 225, 232, 235, 236–37, 238–39; relationship with Gustavo Noboa, 148, 149, 183, 184, 185, 186, 197, 198, 199–200, 202, 208, 226; reputation of, 169, 190–91; salaries of, 92, 180, 191, 192, 222
Armijos, Ana Lucía, 132, 137, 217
Army Finishing School, 173, 175
Arosemena, Carlos Julio, 142
Arroyo del Río, Carlos, 125, 126
Arteaga, Rosalía: on Armed Forces, 96; on corruption, 90; as interim president, 104–7, 232, 233, 244; during 1998 election campaign, 116–17; relationship with Alarcón, 97, 102, 104, 109–10; relationship with Bucaram, 83, 84, 93–94, 96, 98, 99, 100, 101, 102, 103, 105–6, 109–10, 117; relationship with Congress, 96–97, 103–5, 106–7, 242, 244; relationship with Moncayo, 97, 99, 101, 102, 103–4, 105, 106
Aspiazu, Fernando, 129, 133, 141, 142, 146, 155
Atahualpa, 16–18
Atlantic Richfield, 58
AUC (United Self-Defense Forces of Colombia), 145
Austerity measures, 44–46, 48–50, 242; of Alarcón, 111; of Bucaram, 45, 49–50, 85, 92–93, 94, 95, 97–98, 108, 111, 119, 238, 239, 241, 244–45; of Mahuad, 45, 49–50, 111, 116, 118–20, 122, 123–24, 134–40, 143, 157, 159–61, 163, 236, 238–39, 240–41, 244–45; mandated by IMF, 43, 46, 48, 79, 124, 136–37, 139, 143, 159, 160–61, 212, 213–16, 219–20, 222, 246; of Noboa, 216, 219–21, 244; opposed by CONAIE, 46, 47, 135–36, 138, 164, 219, 241, 245. *See also* Government spending
Avenal, Víctor, 228
Ayala Mora, Enrique, 8
Azar, Eduardo, 97
Azuay province, 3, 8, 157

Baca, Raúl, 98
Banco del Progreso, 129, 133, 141, 146, 207
Banco de Préstamos, 129, 130, 133, 134
Banking and finance, 31; corruption in, 124, 129–34, 141–43, 149, 164, 173, 174, 183, 210, 230, 237; Ecuadorean Development Bank, 38; globalization in, 48–49; policies during Mahuad presidency, 130–34, 135, 136, 141–43, 146, 148, 155, 163, 169, 174, 183, 197, 207, 217, 237. *See also* Foreign debt
Baños de Agua Santa, 3–4
Barragan, Gil, 106
Batallas, Hernán, 195
Battle, John, 190
Bayas, Víctor Manuel, 97, 102
Benalcázar, Sebastián de, 7, 8, 18, 21
Better, Pablo, 152
Bolívar, Simón, 24–25
Bolívar province, 3, 8, 73, 78
Bolivia, trade bloc, 189
Bonaparte, Joseph, 23
Bonilla, Edelberto, 74
Borja, Rodrigo: during Bucaram presidency, 93; during Levantamiento Indígena, 166; during Mahuad presidency, 137, 142, 148; during 1998 election campaign, 117, 118, 119; as president, 36, 44–45, 55, 73–74, 78–79, 81, 136, 218
Brady, Nicholas, 139
Brady bonds, 49, 139–40, 160, 224
Brazil, 125–26; trade bloc, 189
Bread and Land of Chimba, 61
Brito, Jorge Luis, 149, 155–56; during Levantamiento Indígena, 170, 177, 181, 183, 184, 186, 194, 195, 199; after Levantamiento Indígena, 208

Brown, Jim, 159
Bucaram, Abdalá: amnesty for Levantamiento Indígena advocated by, 209, 226–27; austerity measures, 45, 49–50, 85, 92–93, 94, 95, 97–98, 108, 111, 119, 238, 239, 241, 244–45; and corruption, 89–90, 92, 94, 98, 112–13, 124, 179, 230, 241–42, 244–45; currency convertibility policies, 87, 94, 98, 107, 245; election campaign, 82–84, 85, 92, 115–16, 119; foreign debt policies, 87; Mahuad compared to, 187, 196, 210, 228, 231, 232–33, 235, 238–39, 241–46; on Mahuad presidency, 157, 237; and modernization, 84, 85, 95, 107–8, 245–46; national strike in opposition to, 94–96; as populist, 45, 82–84, 85, 107, 120; privatization policies, 83–84, 87–89, 107–8, 245; relationship with Alarcón, 95, 97, 101–2, 108; relationship with Armed Forces, 90, 92, 95, 96, 98–101, 102, 105, 107–8, 174, 185, 210, 232, 243; relationship with Arteaga, 83, 84, 93–94, 96, 98, 99, 100, 101, 102, 103, 105–6, 109–10, 117; relationship with Congress, 84, 92, 96–97, 98, 99–101, 179, 232–33, 242, 245; relationship with Moncayo, 90, 92, 95, 96, 98–100, 101, 102, 237, 243; relationship with Álvaro Noboa, 116, 120, 121, 122; relations with Peru under, 89, 90, 92, 94, 108, 127; tax policies, 86–87, 97, 111
Bucaram, Adolfo, 110
Bucaram, Asaad, 82
Bucaram, Santiago, 112
Burbano, Felipe, 192

CAAM (President's Environmental Commission), 59
Cabascango, José María, 69
Cabildos, 66, 72
Calle, Carlos, 224, 225, 226
Camdessus, Michael, 140
Cañar province, 3, 8, 67, 73, 78
Carchi province, 3, 8–9, 13, 57
Carrión, Fernando, 213
Carrión, Francisco, 229
Catholic Church, 22, 23, 26, 29, 30, 52, 53–54, 63, 67, 68
Cavallo, Domingo, 85
CEDOC (Ecuadorean Confederation of Catholic Workers), 76
Center for the Study of Indigenous Education, 78
CEPE (Ecuadorean State Petroleum Corporation), 36
Cevallos, Mario, 208
CFP (Concentration of Popular Forces), 82, 116
Chambers of Commerce and Industry, 103, 106, 206–7, 245
Chamorro, Violeta, 105
Chancosa, Blanca, 67, 69, 72–73, 220
Chapman, Guillermo, 153
Charles III, 22
Chávez, Ángel Polibio, 168
Chávez, Hugo, 174, 180, 228, 237, 243
Chiaquitinta, 18
Chile, 125–26, 174; trade bloc, 189
Chimborazo province, 3, 8, 29, 60, 62, 67, 73, 78
Clare, Gwen C., 188
Clark, Wesley, 90
Clinton, Bill, 148
Coast (Costa), 1, 2, 5, 13, 22, 26, 27, 57, 230, 233; Indians on, 8, 12, 54–55, 70, 72, 73
Cobo, Fausto, 128, 176–77, 181, 183, 186, 191, 194, 195, 208, 228
COICE (Coordinator of Indigenous Organizations of the Coast of Ecuador), 54–55, 73
Colombia, 189; Plan Colombia, 124, 144–46, 218, 229–30, 244; relations with Ecuador, 12, 24, 26, 124, 140–41, 144–46, 216, 228–29, 233; relations with United States, 144–46, 188, 229–30, 244
Colombian Communist Party, 145
Communications Battalion Ruminahui, 175

CONACINIE (National Coordinating Council of the Indigenous Nationalities of Ecuador), 54–55, 68

CONAIE (Confederation of Indigenous Nationalities of Ecuador), 226, 237, 246; Afro-Ecuadorean population according to, 13; and agrarian reform, 65–66, 69, 74, 75, 76, 77, 240, 242; *Ama Llulla, Ama Shua, Ama Quilla* principles of, 77–78; Congress and Supreme Court surrounded/occupied by, 156, 167, 170–71, 172–73, 175–76, 181, 183, 186, 199–200, 225, 235; cultural assertion of, 58–59, 61, 67, 69, 70–72, 77–78, 79, 238, 240, 245; during Bucaram presidency, 93, 94; and education, 78; environmentalism of, 59–60, 79; flag of, 117, 177; and foreign debt, 164, 211; government restructuring advocated by, 70–72, 74, 75, 138, 164, 169, 171–72, 174, 176, 186, 211, 217, 225; Indian population according to, 7–8; and Levantamiento Indígena of 1990, 73–74; moderates and radicals in, 216–18, 245; and modernization, 65–66, 69, 71, 76; multinational and multi-ethnic state advocated by, 70–72, 74, 75; during 1998 election, 122; opposition to austerity measures, 46, 47, 135–36, 138, 147, 164, 219, 223, 241, 245; opposition to corruption, 142, 164, 169, 201, 211; opposition to dollarization, 152, 163–64, 196, 197, 211, 215, 219–20; opposition to Plan Colombia, 146; opposition to privatization, 77, 88, 135–36, 138, 164, 211, 231, 233; organization of, 72–73; origin and composition of, 53, 54–55, 59–62, 66, 67–69, 239–40; as pacifist, 92, 177, 181, 195, 201, 203, 241; Parliament of the People, 155, 164, 166, 171–72, 176, 179, 217; relationship with Armed Forces, 72, 155–57, 165, 166, 167, 168–77, 186–87, 192–94, 199, 201–3, 235; relationship with FUT, 47, 73, 76; relationship with Mahuad, 135–36, 138, 143–44, 156–57, 211; relationship with Gustavo Noboa, 210, 212–13, 216–17; relationship with unions, 47, 61, 73, 76, 79, 221, 223, 245; state decentralization advocated by, 70–71, 138, 174. *See also* Levantamiento Indígena of 2000

CONAM (National Council of Modernization), 206

Concentration of Popular Forces. *See* CFP

Concertaje (debt peonage), 20, 21, 22, 26, 27, 28, 29, 63

Confederation of Ecuadorean Workers. *See* CTE

Confederation of Indigenous Nationalities of Ecuador. *See* CONAIE

Confederation of Peasant and Indian Organizations of Cangahua, 177

CONFENIAE (Confederation of the Indigenous People of the Ecuadorian Amazon), 54, 68, 69, 70, 240

Congress: Alarcón chosen interim president by, 98, 99–100, 101–3, 104–5, 106–7, 108–10; and amnesty for Levantamiento Indígena, 209; constitutional relationship to executive, 47, 104, 105, 106, 110, 115–16, 123; corruption in, 110, 112, 211; Economic Transformation Law passed by, 214; and election procedure, 115–16; and Law to Promote Investment and Citizen Participation, 231–33; Gustavo Noboa proclaimed president by, 205; and political fragmentation, 115–16, 118–19, 161; relationship with Arteaga, 96–97, 103–5, 106–7, 242, 244; relationship with Bucaram, 84, 92, 96–97, 98, 99–101, 179, 232–33, 242, 245; relationship with Mahuad, 123, 124, 137, 138, 140,

143, 245; surrounded/occupied during Levantamiento Indígena, 156, 167, 170–71, 172–73, 175–76, 181, 183, 186, 187, 199–200, 225, 235
Conservative Party. *See* CP
Constitution of 1998, 77–78
Continental Indian Congress of 1990, 74
Coordinated Social Movements. *See* CSM
Coordinator of Indigenous Organizations of the Coast of Ecuador. *See* COICE
Corruption: and Alarcón, 179, 241–42; in banking and finance, 124, 129–34, 141–43, 149, 164, 173, 174, 183, 210, 230, 237; during Bucaram presidency, 89–90, 92, 94, 98, 112–13, 124, 179, 230, 241–42, 244–45; in Congress, 110, 112, 211; during Mahuad presidency, 141–42, 146, 148, 154–55, 169, 173, 175, 180–81, 183, 207, 230, 237, 239, 241–42, 244–45; extent of, 89–90, 108, 112–13, 169, 179, 190–91, 211, 237, 241–42, 246, 247; in Guayaquil customs house, 90, 113, 149, 175; and Gustavo Noboa, 207, 212; opposed by Armed Forces, 173, 175, 179–80, 181, 191, 192, 227, 237; opposed by CONAIE, 142, 164, 169, 201, 211; as political issue, 77, 108, 117, 120, 121; and Solórzano, 179
Cortés, Hernán, 17
Cotopaxi Indigenous and Peasant Movement, 70
Cotopaxi province, 3, 8, 63, 70, 73, 78, 138
CP (Conservative Party), 108, 116
CPM (Women's Political Movement), 93
Crespo, Ramiro, 153
Cristiani, Alfredo, 129
Cruz, Pedro de la, 67
CSM (Coordinated Social Movements), 70, 76, 79, 93, 94, 165, 175, 215, 217, 219, 246. *See also* Saltos, Napoleón
CTE (Confederation of Ecuadorean Workers), 61, 76
Cuadra, José de la, 61
Cuenca, 83, 184
Curicama Guamán, Mariano, 219

Dahik, Alberto, 124, 179
Daquilema, Fernando, 28–29, 60
Darwin, Charles, 5
Dávalos, Mauricio, 222
Debt peonage, 20, 21, 22, 26, 27, 28, 29, 63
Decentralization of the state, 70–71, 138, 174
Deforestation, 57, 58–59, 68
Democratic Left Party. *See* ID
Deposit Guarantee Agency. *See* AGD
Development Fund for Indian Communities, 222
Díaz, César, 172, 175
Díaz, Rui, 18
DINE, 36
Dominguel Bucheli, Oswaldo, 226
Donoso Morán, Fernando, 188, 226
DP (Popular Democracy Party), 83, 98, 108, 116, 118, 132, 161, 205, 207, 209, 226, 227, 230. *See also* Hurtado, Osvaldo; Mahuad, Jamil
Dunbar, Ron, 140
Durán Abad, César, 100, 101
Durán Arcentales, Guillermo, 39
Durán Ballén, Sixto: during Noboa presidency, 212; as president, 45, 49, 57, 75, 76, 79, 81, 83, 88, 92, 108, 126; relations with Peru under, 126

Ecochicas, 59, 75
Ecological Science, 59
Economic conditions: distribution of wealth, 46–47, 62, 77, 159, 240, 247; electricity rates, 38, 42, 43, 45, 86, 97, 111, 119, 220; exports, 24, 26, 27, 30, 33, 34–35, 37, 39, 41; fishing, 65; foreign investment, 41, 48–49, 85, 95, 103, 214; and globalization, 40, 47–48; import-substitution policies, 37, 39, 41, 49, 63, 240;

Economic conditions (*continued*) inflation, 40, 42, 44, 45, 46, 85, 111, 124, 134–35, 137, 149–50, 151, 157–58, 160, 216, 221, 223; interest rates, 40, 42–43, 111, 223; minimum wage, 44, 121, 158, 214, 222; poverty rate, 45–46, 70, 78, 158, 159, 214, 222, 227, 240–41, 247; prices of cooking fuel, 38, 42, 86, 119; prices of gasoline, 38, 42, 43, 44, 86, 93, 97, 124, 136, 137–38, 139, 220, 221, 222; state-ownership, 36–37, 43, 48, 49, 81–82, 87–89, 111, 214, 215, 222, 223, 231, 245; textile manufacturing, 22; unemployment, 44, 45, 158, 159, 216. *See also* Agrarian reform; Agriculture; Foreign debt; Government spending; Neoliberalism; Oil; Privatization; Taxes
Economic Transformation Law (Ley Trole I), 214, 215–16
Ecuador: Afro-Ecuadoreans in, 7, 13, 27, 28, 71, 77, 159; Cholos/ Mestizos in, 6, 7, 8, 12, 30, 62–63, 69; climate, 5–6; geography of, 1–5; under Incas, 3, 8, 9, 15–18, 21, 78; independence movement in, 23–25; population of, 6–8, 12–13, 63; racial classification in, 7–8, 12–13; relations with Colombia, 12, 24, 26, 124, 140–41, 144–46, 216, 228–29, 233; relations with Peru, 43, 59, 89, 90, 92, 94, 98, 102, 108, 113, 124–28, 172, 180, 191; relations with United States, 30, 33, 62, 63, 90, 96, 99, 100, 102–3, 107, 125–26, 140, 143, 144–46, 148, 153, 185, 187, 188–89, 207, 212, 214–15, 218, 229, 242, 243–44, 246; under Spanish rule, 1, 6, 7, 8, 16–24, 52; whites in, 7, 12–13. *See also* Armed Forces; Congress; Indians; Supreme Court
Ecuadorean Communist Party. *See* PCE
Ecuadorean Confederation of Catholic Workers. *See* CEDOC
Ecuadorean Federation of Evangelical Indians. *See* FEINE
Ecuadorean Indigenous Federation. *See* FEI
Ecuadorean Revolutionary Socialist Party. *See* PSRE
Ecuadorean Roldista Party. *See* PRE
Ecuadorean Socialist Party. *See* PSE
Ecuadorean State Petroleum Corporation. *See* CEPE
Ecuadorean Workers Confederation, 61
ECUARUNARI (Ecuador Ruñacunapac Riccharimui), 54–55, 67, 68, 69, 70, 72, 73, 144, 167, 177, 179, 210, 217–18, 240
Education, 20, 78, 119–20, 121, 158
Ehlers, Freddy, 83, 84, 117, 118, 138
Election procedures, 115–16
El Inca, 29, 61
El Niño, 5–6, 40, 44, 116, 118, 121, 124, 160, 205
ELN (National Liberation Army), 145
El Oro province, 8, 12, 13, 125
Eloy Alfaro Military Air Base, 144
El Salvador, 129
EMETEL, 88–89, 97
Encomienda system, 20, 21
Environment: Ecochicas and, 59, 75; effects of oil on, 56, 57–60, 75, 177; environmental groups, 53, 57–58, 59, 68, 69, 244; environmentalism among Indians, 54, 59–60
Erazo, Luis, 208
Escuela Militar Eloy Alfaro, 181, 233
Esmeraldas province, 12, 13, 33, 67, 71
Espinoza, Nicolás, 94–95
Eurobonds, 224

Falconi, Juan, 134, 151, 230
FARC (Revolutionary Armed Forces of Colombia), 145, 228–29
FARE (Revolutionary Armed Forces of Ecuador), 229

Febres Cordero, León, 7, 227, 232, 233; on Armed Forces, 192; during Bucaram presidency, 93, 95, 103, 108; during Levantamiento Indígena, 188; after Levantamiento Indígena, 209; during Mahuad presidency, 138, 148, 157, 188; as president, 44, 78, 81, 87, 88, 120, 136
Federation of Chambers of Agriculture of Ecuador, 147
Federation of Indians and Campesinos of Imbabura. *See* FICI
Federation of Indigenous Organizations of the Napo. *See* FOIN
Federation of Saraguros of Zamora Chinchipe, 179
Federation of Shuar Centers, 53–54, 67–68
FEI (Ecuadorean Indigenous Federation), 61
FEINE (Ecuadorean Federation of Evangelical Indians), 53, 67, 218, 220
FENOC (National Federation of Farmers' Organizations), 62–63, 66–67
FENOCIN (National Federation of Rural Campesinos, Indians, and Negroes), 67
Ferdinand VII, 23, 24
FICI (Federation of Indians and Campesinos of Imbabura), 70
Fierro, Francisco, 228
Filanbanco, 129–30, 132, 133–34, 230
Fischer, Stanley, 213–14
FLOPEC, 26
Flores, Juan José, 25, 26, 28
Flores Zapata, Gustavo, 113
FOIN (Federation of Indigenous Organizations of the Napo), 54
Forbes, Steve, 121–22
Foreign debt, 39, 40, 42–43, 48–49, 81, 85, 116; Brady bonds, 49, 139–40, 160, 224; and CONAIE, 164, 211; government spending on, 110, 158–59, 160, 233–34; and IMF, 43, 45, 46, 48, 79, 84, 124, 136–37, 139, 140, 143, 150, 159, 160–61, 212, 215–16, 219–20, 222, 224, 246; policies during Alarcón presidency, 108, 110–11; policies during Bucaram presidency, 87; policies during Mahuad presidency, 119, 121, 124, 127, 135, 136–38, 139–40, 143, 150, 158–59, 160–61, 245, 246; policies during Noboa presidency, 212, 213–14, 215–16, 219–20, 222, 223, 231, 233–34. *See also* Banking and finance
FP (Popular Front), 70, 220, 229
France, criticism of January 2000 coup, 187
Franks, Jeffrey, 130
FRA (Radical Alfarista Front), 97, 98, 108, 109–10, 116, 226
Free Land of Moyurco, 61
Fuertes, Homero, 112
Fujimori, Alberto, 126, 127, 135, 230
FUT (United Workers Front), 47, 73, 76, 93, 98–99

Galápagos Islands, 1, 5, 12
Gallardo, Jorge, 224
Gallardo, José, 141, 147, 149, 151, 192–93, 200, 202, 224–25, 235, 236; relationship with Mahuad, 154–55, 163
Gamarra, Eduardo, 79
García, Patricia, 113
García Moreno, Gabriel, 26
Gaviria, César, 189, 200, 203, 212
GCP (Popular Fighters Group), 141
Gende, Ángel, 72
Germán Carrasco, Cabo, 228
Globalization, 40, 47–49, 233–34
González, Susana, 232, 233
González Alvear, Raúl, 39
Goodyear, Charles, 51
Gortaire, Víctor, 228
Government spending, 110–12, 119–20, 136, 137, 180, 214, 236; for Armed Forces, 26, 36, 90, 92, 127, 174, 180, 191, 192, 222, 246; on foreign debt, 110, 158–59, 160, 233–34; vs. government revenue, 34, 35, 36–37, 39–42,

Government spending (*continued*) 45, 55, 75, 239, 240–41; relationship to oil, 35, 36, 37, 38–40, 42, 43, 44, 45, 46, 56–57, 75, 81, 239, 240–41; for subsidies, 36, 38–39, 42, 43, 45, 49, 85–86, 93, 111, 116, 119, 120–21, 123, 124, 132, 135, 143, 174, 216, 219, 220, 221–22, 239, 240–41, 244–45. *See also* Austerity measures; Taxes
Gran Colombia, 24–25
Guatemala, 129
Guayaquil, 2, 22–23, 24, 26, 44, 81, 136, 147, 152, 184, 188; and Bucaram, 82, 84, 89, 101, 102; customs house corruption in, 90, 113, 149, 175
Guayas province, 8, 12, 13, 81, 82
Gueiler Tejada, Lydia, 105
Gutiérrez, Lucio, 142; on corruption, 237; during Levantamiento Indígena, 155, 166, 167, 173–76, 176–77, 179–80, 181, 183–84, 186, 187, 192, 194–95, 196, 197, 199, 200, 201, 203, 235; after Levantamiento Indígena, 201, 208, 225, 227–28, 237, 243; and Mendoza, 166, 179, 183–84, 186, 189, 192, 194–95, 199, 200, 201, 203; on neoliberalism, 237; relationship with Vargas, 173, 176; on U.S. opposition to Junta of National Salvation, 189
Guzmán, Jorge, 221–22

Hacienda (huasipungo) system, 20, 28, 29, 61, 63, 64, 66
Hacienda Leito, 29–30
Hanke, Steve, 153
Hassaurek, Friedrich, 27, 28
Hernández Peñaherrera, Luis, 181
Heroes of the Cenepa Engineering Brigade, 172, 173, 175, 176
Hidalgo Villacís, Carlos, 113
Highlands (Sierra), 1, 2–4, 11, 13, 22, 23, 26, 27, 29, 57, 58–59, 61, 64, 65, 66, 230, 233; Indians in, 2, 3, 8–9, 54–55, 67, 68, 69–70, 72, 73, 74, 240
Hinojosa, Marco, 100, 113
Huaorani, 4, 8, 10, 11, 51–53, 55–56, 57–58, 59, 73–75, 79. *See also* Indians
Huáscar, 16–18
Huayna Cápac, 16
Huerta, Francisco, 62, 149, 207, 213, 217
Human rights groups, 53, 57–58, 59, 68, 69, 244
Hurtado Larrea, Osvaldo, 7; during Bucaram presidency, 89, 93; during Levantamiento Indígena, 187–88; during Mahuad presidency, 132, 137, 148, 185; during Noboa presidency, 212, 215; as president, 43–44, 78, 81, 118, 129–30, 136

Icaza, Jorge, 61
ICCI (Scientific Institute of Indian Cultures), 78
ID (Democratic Left Party), 98, 116, 117, 123, 226, 230–31. *See also* Borja, Rodrigo
Imbabura province, 3, 8, 9, 13, 57, 67, 70, 71, 73
IMF (International Monetary Fund), 45, 84, 87, 130, 140, 150, 224; austerity measures mandated by, 43, 46, 48, 79, 124, 136–37, 139, 143, 159, 160–61, 212, 213–16, 219–20, 222, 246
Immigrants, to Oriente, 57, 58–59, 68
Import-substitution policies, 37, 39, 41, 49, 63, 240
Incas, 3, 8, 9, 15–18, 21, 78
Indian Fund, 213
Indians: Achuar, 4, 8, 10–11, 73, 126; and agrarian reform, 63–67, 68, 69, 74, 75, 76, 77, 78; agriculture among, 3, 11, 53, 63–67, 70; Arda, 11; in Armed Forces, 191; attitudes of non-Indians toward, 12, 13, 19–20, 25, 27, 28–29, 51–52, 60, 62, 67, 68–69, 71, 74, 218–19; Awá, 8, 12, 73; Bolona, 11; Bracamoro, 11; Cañari, 3, 8, 18, 73; Canelos, 9; Caranqui, 3, 8; Cayambi, 3, 8; Cayapa, 12, 52; Chachis, 8, 12, 73; Chibuelos, 3, 8, 73; Chirino,

11; on the coast, 8, 12, 54–55, 70, 72, 73; Cofánes, 4, 8, 11, 52, 56, 73; communal lands of, 25; diseases among, 20, 52; effects of oil development on, 10, 11, 51–53, 56, 57–60, 68, 69, 73–75, 239–40; effects of outsiders on, 4, 10, 11–12, 51–53, 57, 58–59; environmentalism among, 54, 59–60; Épera, 12; in the Galápagos, 12; government policies toward, 51–53, 55, 68–69, 76, 77, 78–79; in the highlands, 2, 3, 8–9, 54–55, 67, 68, 69–70, 72, 73, 74, 240; Huancavilcas, 12, 73; Huaorani, 4, 8, 10, 11, 51–53, 55–56, 57–58, 59, 73–75, 79; Karamkis, 73; Kayampis, 73; Mantas, 12, 73; Natabuelas, 73; in the Oriente/Amazon, 4, 8, 9–12, 51–60, 67–68, 69–70, 72, 73–75, 79, 239–40; Oriente Quichuas, 4, 8, 9–10, 52; Otavalenos/Otavalos, 3, 8, 9, 67, 73; Pan American Highway blocked by, 4, 74, 94, 241; Panzaleos, 3, 8, 73; population, 7–8, 11, 60; Puruháes, 3, 8, 73; Quichuas, 8, 9; Quijos, 9; Quishapinchas, 73; Quitu, 3, 8, 73; relations with Spaniards, 19–21; Salasacans, 3, 8, 73; Saraguros, 3, 8, 9, 73, 179; Secoya, 4, 8, 11, 52, 73; Shuar/Jívaros, 4, 8, 10, 11, 52, 53–54, 67–68, 73, 126; Siona, 4, 8, 11, 73; socioeconomic position of, 6, 19–20, 26–28, 62, 63, 68, 69, 70, 78, 159, 216–17; Tete, 11; tribute from, 20, 21–22, 26–27; Tsachila, 8, 12, 71, 73; Tugua, 3, 8; Warankas, 3, 8, 73; Záparos, 4, 11, 73. *See also* CONAIE
Indigenous Chamber of Commerce, 219
Inflation. *See* Economic conditions, inflation
Inter-American Development Bank, 43, 65, 127, 130, 140, 213, 215
International Monetary Fund. *See* IMF
Inti Raymi, 70
Irigoyen, Ricardo, 195, 226
Isais, William and Roberto, 129, 130, 133–34
Iturralde, Diego, 168–69
Iturralde, Pablo, 175
Iza, Leonidas, 69

Jaramillo Avarado, Pío, 29, 60
Jarrín, Oswaldo, 188
Jesuit Order, 22
Jívaros. *See* Shuar/Jívaros
J. P. Morgan, 224
Junta of Economic Planning and Coordination, 38
Junta of National Salvation: composition of, 176–77, 179–80, 186, 194–96, 197–98, 203, 224, 225; opposition to, 187–90. *See also* Gutiérrez, Lucio; Levantamiento Indígena of 2000; Mendoza, Carlos; Solórzano, Carlos; Vargas, Antonio

Kimerling, Judith, 57–58

Labaca, Archbishop Alejandro, 52
Lacano Palacios, Mario, 181
Lalama, Gustavo, 177, 181, 183, 184, 186, 194, 195, 208
Landes, Nicolás, 134
Land reform. *See* Agrarian reform
Larrea, Salomón, 154–55
Larreátegui, Carlos, 185
Lascano, Mario, 228
Lascano Yañez, José, 224, 225, 226
Las Casas, Bartolomé de, 19
Law of Communes (Ley de Comunas), 66
Law of Financial Institutions, 129
Law of National Security, 190
Law of Political Parties, 141
Law of the Dead Hands (Ley de Manos Muertas), 29, 61
Laws of the Indies, 20
Law to Promote Investment and Citizen Participation, 231–32
León Mera, Juan, 60
Leoro Franco, Luis, 39
Levantamiento Indígena of 1990, 73–74

Levantamiento Indígena of 2000, 80, 147, 155–57, 241; Álvarez during, 165, 167, 168, 170, 184, 185, 186; amnesty for participants in, 216, 218, 223, 224–28, 233; Armed Forces during, 155–56, 163, 165, 167–77, 179–80, 181, 183–85, 186, 188, 192–93, 209–10, 227–28, 238–39, 243; Brito during, 170, 177, 181, 183, 184, 186, 194, 195, 199; Gutiérrez during, 155, 166, 167, 173–76, 176–77, 179–80, 181, 183–84, 186, 187, 192, 194–95, 196, 197, 199, 200, 201, 203, 235; Mahuad during, 165, 167–68, 184–86, 238; Mendoza during, 156, 164, 165, 166–71, 181, 183–85, 186–87, 192–200, 201, 202–3, 210, 224, 225, 235, 236–37; Moncayo during, 181, 209, 224, 243; Gustavo Noboa during, 183, 184, 185, 186, 187, 197, 198, 199–200, 202, 235; Quishpe during, 167, 175, 179, 194, 195; regional military commanders during, 184, 188, 197, 200, 202, 236; Sandoval during, 166, 167, 170–71, 176, 183, 184, 185, 192, 194–95, 197, 198, 199–200, 202–3, 224, 225, 235, 236; Vargas during, 163, 164, 166, 168, 172–73, 176, 186, 187, 193–94, 197, 201–2, 235. *See also* CONAIE
Liberal Party, 63
Lima Garzón, María Eugenia, 117, 118
Lluco, Miguel, 69
Logistical Support Brigade Number 25, 175
Loja province, 3, 8, 9, 59, 73, 125, 184
Loor, Jorge, 198
López, Gonzalo, 229
López, Patricio, 113
López, Raúl (Bishop of Laracunga), 63, 79, 159
Los Ríos province, 12
Luna, Milton, 115

Macas, Louis, 69, 73–74, 76, 92, 157, 217; on agrarian reform, 65–66; on austerity measures, 46; on environmental destruction in Oriente, 59–60; on IMF/World Bank, 46; on modernization, 65–66, 71; on state decentralization, 71
Mahuad, Eduardo, 141, 146
Mahuad, Jamil: austerity measures, 45, 49–50, 111, 116, 118–20, 122, 123–24, 134–40, 143, 157, 159–61, 163, 236, 238–39, 240–41, 244–45; banking policies, 130–34, 135, 136, 141–43, 146, 148, 155, 163, 169, 174, 183, 197, 207, 217, 237; Bucaram compared to, 187, 196, 210, 228, 231, 232–33, 235, 237, 238–39, 241–46; during Bucaram presidency, 94, 95, 103, 104; and corruption, 141–42, 146, 148, 154–55, 169, 173, 175, 180–81, 183, 207, 230, 237, 239, 241–42, 244–45; dollarization policies, 124, 148–54, 163–64, 171, 174, 189, 196, 197, 207, 237, 245; education policies, 119–20, 158; election campaign of, 116, 118–20, 122–23, 141; foreign debt policies, 119, 121, 124, 127, 135, 136–38, 139–40, 143, 150, 158–59, 160–61, 245, 246; during Levantamiento Indígena, 165, 167–68, 184–86, 238; as mayor of Quito, 94, 95, 103, 108, 116, 118; monetary policies, 124, 132, 148–54, 157–58, 163–64, 171, 174, 189, 196, 197, 207, 237, 245; privatization policies, 119, 138, 140, 189, 245; relationship with Armed Forces, 127–28, 132–33, 137, 138, 139–40, 142, 146–49, 151, 154, 163, 165, 166–71, 173–75, 184–85, 186, 191–92, 199–200, 202, 208, 210, 225, 232, 235, 236–37, 238–39; relationship with CONAIE, 135–36, 138, 143–44, 156–57, 211; relationship with Congress, 123, 124, 137, 138, 140, 143, 245; relationship with Gallardo, 154–55, 163; relationship with Mendoza, 127, 132–33, 135, 137, 138, 139–

40, 141–42, 147–48, 149, 151, 155, 156, 166, 167–71, 184–85, 199–200, 202–3, 225, 236–37; relationship with Gustavo Noboa, 132, 137, 141, 146, 148, 156, 185, 203, 206, 211, 218, 238; relations with Peru under, 124–28, 180; reputation of, 207; subsidy policies, 116, 119, 123, 124, 132, 135, 136, 137–38, 139; tax policies, 116, 118, 120, 122, 123, 124, 131, 135, 136, 163, 245
Maiguashca, Segunda B., 61
Manabí province, 12, 144
Manta, 144–46, 218, 229–30
Martínez, Cristóbal, 100
Martínez, Esperanza, 59
Mendoza, Carlos: on banking crisis, 132–33; and Gutiérrez, 166, 179, 183–84, 186, 189, 192, 194–95, 199, 200, 201, 203; during Levantamiento Indígena, 156, 164, 165, 166–71, 181, 183–85, 186–87, 192–200, 201, 202–3, 210, 225, 235, 236–37; after Levantamiento Indígena, 208–9, 236–37; as member of Junta of National Salvation, 194–98, 199, 200, 201, 203, 224, 225; on peace treaty with Peru, 127; relationship with Cobo, 176; relationship with Mahuad, 127, 132–33, 135, 137, 138, 139–40, 141–42, 147–48, 149, 151, 155, 156, 166, 167–71, 184–85, 199–200, 202–3, 225, 236–37; relationship with Sandoval, 166, 167, 170–71, 183, 194–95, 197, 199–200, 202–3, 224, 235, 236; U.S. pressure on, 189
Menem, Carlos, 129
Mercosur trade bloc, 189
Merino, Wilson, 109–10
Mexico, 17, 93, 129, 140
Military Polytechnic School, 172, 173, 175, 176, 183, 192
Miño, Marco, 173
MIRA (National Alliance Party), 116–17
Missionaries, 51–53, 67, 244
Mita system, 20, 21, 29
Modesto Paredes, Ángel, 60–61
Moeller, Heinz, 137, 207, 213, 215
Molano, Walter, 152–53
Moncayo, Carlos, 185
Moncayo, Paco: during Levantamiento Indígena, 181, 209, 224, 243; after Levantamiento Indígena, 227, 237, 240, 243; popularity of, 117, 126, 155, 173; relationship with Alarcón, 99, 102–3; relationship with Arteaga, 97, 99, 101, 102, 103–4, 105, 106; relationship with Borja, 117; relationship with Bucaram, 90, 92, 95, 96, 98–100, 101, 102, 237, 243
Moncayo Gallegos, Carlos, 170
Monetary policies, 130, 157; during Bucaram presidency, 87, 94, 98, 107; convertibility of the sucre, 87, 94, 98, 107, 132, 245; dollarization, 124, 148–54, 163–64, 171, 174, 189, 196, 197, 207, 211–12, 213–15, 219, 237, 245; exchange rates, 124, 149–50, 197, 213–14; during Mahuad presidency, 124, 148–54, 163–64, 171, 174, 189, 196, 197, 207, 237, 245
Monsalve Pozo, Luis, 60–61
Monteverde, Enrique, 194–95, 226
Morga, Antonio de, 19
Morona Santiago province, 4, 10, 53
MPD (Popular Democratic Movement), 98, 116, 117, 226
Murillo, Marco, 67, 218, 220

Napa, Calixto, 72
Napoleon I, 23
Napo province, 4, 9, 54, 57, 73, 177
Narváez, Norton, 208, 226
National Accord Forum, 136
National Alliance Party. *See* MIRA
National Confederation of Afro–Ecuadoreans, 13
National Coordinating Council of the Indigenous Nationalities of Ecuador. *See* CONACINIE
National Council of Modernization. *See* CONAM

National Federation of Farmers' Organizations. *See* FENOC
National Federation of Rural Campesinos, Indians, and Negroes. *See* FENOCIN
National Institute for Agricultural Development (INDA), 173
National Liberation Army. *See* ELN
National Peasant Council, 198
National Police, 130
National Union of Educators, 220
Natural gas, 41, 42
Nature Foundation, 59
Nebot, Jaime, 81–82, 83, 84, 108, 109, 117–18, 119, 122, 135, 226, 232
Nebot Velasco, Jaime, 82
Neira, Xavier, 232, 233
Neoliberalism, 40, 44–45, 48–50, 71, 122, 129, 188, 237, 244, 246. *See also* Austerity measures; Government spending; Privatization
New Country Party. *See* NP
Noboa, Álvaro, 83, 100, 118, 119, 120–23; as populist, 120, 122; relationship with Bucaram, 116, 120, 121, 122
Noboa, Diego, 205
Noboa, Gustavo, 205–10; amnesty policies, 49, 216, 223, 224–28, 233; austerity measures, 216, 219–21, 244; banking policies, 207, 211, 216; and corruption, 207, 212; dollarization policies, 211–12, 213–15, 216; foreign debt policies, 212, 213–14, 215–16, 219–20, 222, 223, 231, 233–34; on Indian movement/Levantamiento Indígena, 62, 156, 209, 215, 218–19, 237–38; during Levantamiento Indígena, 183, 184, 185, 186, 187, 197, 198, 199–200, 202, 235; privatization policies, 214, 215, 216, 220, 222, 223, 231–32, 233; relationship with Armed Forces, 148, 149, 183, 184, 185, 186, 197, 198, 199–200, 202, 208, 226; relationship with CONAIE, 210, 212–13, 216–17; relationship with Mahuad, 132, 137, 141, 146, 148, 156, 185, 203, 206, 211, 218, 238; reputation of, 207
Noboa, Luis, 106
Noboa, Ricardo, 205, 206, 231
Noboa Naranjo, Luis, 120
Nongovernmental organizations (NGOs), 68, 135
NP (New Country Party), 83, 116, 117, 123, 230
Nueva Loja (Lago Agrio), 51

OAS (Organization of American States), 187, 189, 200, 203, 212, 242, 243–44
Oil: and Armed Forces, 36; CEPE (Ecuadorean State Petroleum Corporation), 36; discovery and exploration, 2, 10, 51–53, 55–60; effects on Ecuador, 10, 11, 35–39, 115, 244; effects on environment, 56, 57–60, 75, 177; effects on Indians, 10, 11, 51–53, 56, 57–60, 68, 69, 73–75, 239–40; government revenue from, 34, 35, 36–37, 40–42, 45, 55, 75, 160, 165, 239, 240–41; Petroecuador, 36, 41, 56, 113, 165, 231; price of, 34–35, 40–43, 44, 45, 47, 48, 55, 56–57, 73–74, 75, 81, 85, 111, 123–24, 160, 161, 216, 239, 245; relationship to government spending, 35, 36, 37, 38–40, 42, 43, 44, 45, 46, 56–57, 75, 81, 239, 240–41; and taxes, 35; and tourism, 55, 56; Trans-Ecuadorean Pipe Line (SOTE), 33–34, 40, 44, 56, 57, 140–41, 216
OPEC (Organization of Petroleum Exporting Countries), 34–35, 40, 41, 42–43
OPIP (Organization of Indigenous People of Pastaza), 54, 55, 70, 177
Orden, Luis, 134
Orellana province, 9, 10–11, 13, 51, 55–56, 73
Organic Law of the Armed Forces, 190
Organization of American States. *See* OAS

Organization of Indigenous People of Pastaza. *See* OPIP
Organization of Petroleum Exporting Countries. *See* OPEC
Oriente/Amazon, 1–2, 5, 33–34, 66; immigrants to, 57, 58–59, 68; Indians in, 4, 8, 9–12, 51–60, 67–68, 69–70, 72, 73–75, 79, 239–40; oil development in, 2, 10, 11, 51–53, 55–60, 68, 69, 73–75, 239–40
Ortíz, Benjamín, 135, 145, 235
Oviedo, Gonzalo Fernándz, 19

Pacari, Nina, 69, 74, 76
Pachakutik (New Awakening and Revolutionary Change) Party, 74, 76–77, 79, 98, 122, 123, 196, 217, 230–31; amnesty for Levantamiento Indígena supported by, 209, 226; and NP, 83, 116, 117
Pachano, Abelardo, 86
Pachano, Simón, 139
Pan American Highway, 4, 74, 94, 241
Paraguay, trade bloc, 189
Paris Club, 43, 137, 140, 143
Pastaza province, 4, 9, 10–11, 17, 51, 53, 54, 55, 70, 73, 184
Pastrana, Andrés, 229
Patinio, Claudio, 86
Patriotic Front, 93, 220–21
Paz, Rodrigo, 83, 108, 146–47
PCE (Ecuadorean Communist Party), 61, 76
Peñafiel, Alejandro, 129, 130, 134
Peñafiel, Ruth, 72
Peñaherrera, Blasco, 171
Pérez, Álvaro, 97
Pérez, Ignacio, 74
Perón, Isabel, 105
Peru, relations between Ecuador and, 43, 59, 98, 102, 113, 172, 191; during Bucaram presidency, 89, 90, 92, 94, 108, 127; during Mahuad presidency, 124–28, 180
Petroecuador, 36, 41, 56, 113, 165, 231
Petroleum. *See* Oil
Piaguaje, Anibal, 72
Pichincha province, 3, 8, 12, 13, 29, 57, 61, 62, 67, 73, 108, 118
Pickering, Thomas R., 212
Pico, Galo, 170, 208–9, 217
Pino, Jimmy, 175
Pinto, Gustavo, 94, 158, 219
Pinto Rubianes, Pedro, 205
Pizarro, Francisco, 17, 18
Plan Colombia, 124, 144–46, 218, 229–30, 244
Plaza Lasso, Galo, 12, 28, 33, 36
Pons, Juan José, 207, 212, 226
Popular Democracy Party. *See* DP
Popular Democratic Movement. *See* MPD
Popular Fighters Group. *See* GCP
Popular Front. *See* FP
Popular Fund, 111
Poraño, Monsignor Leonidas, 39
Poveda Burbano, Alfredo, 39
Poverty, 124, 234, 246; Armed Forces and the poor, 169, 190, 191–92, 240; Bucaram and the poor, 82–84; and government subsidies, 221–22, 223; among Indians, 62, 63, 68, 70, 78, 79, 159; poverty rate, 45–46, 70, 78, 158, 159, 214, 222, 227, 240–41, 247
PRE (Ecuadorean Roldista Party), 82, 84, 108, 110, 212, 230–31, 233; and amnesty for Levantamiento Indígena, 209, 226–27. *See also* Bucaram, Abdalá; Noboa, Álvaro
President's Environmental Commission. *See* CAAM
Privatization, 43, 49, 71, 81–82, 111, 189, 206; opposed by Armed Forces, 88; opposed by CONAIE, 77, 88, 135–36, 138, 164, 211, 231, 233; policies during Bucaram presidency, 83–84, 87–89, 107–8, 245; policies during Mahuad presidency, 119, 138, 140, 189, 245; policies during Noboa presidency, 214, 215, 216, 220, 222, 223, 231–32, 233
Proaño, Leonardo, 30
Proaño, Luis, 209–10

Proaño Maya, Marco, 96
PSC (Social Christian Party), 83, 98, 108, 116, 122, 123, 131, 196, 205, 230–31, 232, 233; amnesty for Levantamiento Indígena opposed by, 209, 226, 227. *See also* Febres Cordero, León; Nebot, Jaime
PSE (Ecuadorean Socialist Party), 75–76, 117, 123
PSRE (Ecuadorean Revolutionary Socialist Party), 76
PUR (United Republican Party), 45

Quevado, Hugo, 233
Quimbo, José, 79
Quiroz, Hernán, 100
Quishpe, Salvador, 69, 152, 210, 217–18, 220, 223; during Levantamiento Indígena, 167, 175, 179, 194, 195
Quisquis, 18
Quito, 1, 3, 18, 23–24, 26, 37, 63–64, 108, 112

Radical Alfarista Front. *See* FRA
Rainbow (environmental group), 59
Reagan, Ronald, 44
Reed, John S., 153
Remache, Estuardo, 72, 218
Revolutionary Armed Forces of Colombia. *See* FARC
Revolutionary Armed Forces of Ecuador. *See* FARE
Revolutionary Túpac Amaru Movement, 127
Rijalba, Alfredo, 142
Roca, Saona, 229
Rocafuerte, Vicente, 26
Rodríguez Lara, Guillermo, 35–36, 38, 39, 67, 192
Rojas, Rosendo, 122
Roldós, Jaime, 39, 43, 69, 78, 81, 82, 85, 93
Roldós, León, 153
Romero, Peter, 143, 189
Romero Castañeda, María de Lourdes, 122
Rubio Orbe, Gonzalo, 61
Rumiñahui, 17, 18, 71

Salem Kronfle, Miguel, 97, 98, 113
Salesian Order, 52, 53–54, 67, 68
Salinas de Gortari, Carlos, 129
Salomon Smith Barney, 224
Saltos, Napoleón, 70, 136, 164, 223
Sandoval, Telmo: during Levantamiento Indígena, 166, 167, 170–71, 176, 183, 184, 185, 192, 194–95, 197, 198, 199–200, 202–3, 224, 225, 235, 236; after Levantamiento Indígena, 208, 226; relationship with Mendoza, 166, 167, 170–71, 183, 194–95, 197, 199–200, 202–3, 224, 235, 236
San Martín, José, 24
Santos, Marcelo, 213
Saona Roca, Miguel, 226
Saudi Arabia, 87
Scientific Institute of Indian Cultures. *See* ICCI
Shining Path, 127
Shuar/Jívaros, 4, 8, 10, 11, 52, 53–54, 67–68, 73, 126. *See also* Indians
Sicouret, Víctor Hugo, 232
Silva, Gladys, 215
Social Christian Party. *See* PSC
Social Security Institute, 130
Solórzano, Carlos, 117, 164, 176, 179, 186, 187, 188, 194, 195–96, 197, 201, 209
Spanish rule, 1, 6, 7, 8, 16–24, 52
Standard Oil, 33
Strikes, national, 47, 73, 76, 93, 94–96, 124, 135, 136, 139, 223
Suárez, Víctor, 113
Suasnavas, Gonzalo, 208
Subsidies. *See* Government spending, for subsidies
Sucre, Antonio José de, 24
Sucumbíos province, 9, 11, 13, 33, 51, 56, 73, 144, 229
Summer Institute of Linguistics (Wycliffe Bible Translators), 51–53
Summers, Larry, 153
Sun Celebration, 70
Supreme Court, 113, 169, 209, 211, 217, 242

Tamayo, José Luis, 30
TAME, 36
Tapui, Cristóbal, 69
Taukowash, Miguel, 69
Taxes: during Alarcón presidency, 111; attitudes toward, 26–27, 43, 86–87, 118–19, 161, 236, 239, 240; during Bucaram presidency, 86–87, 97; as election issue, 116, 117, 118, 120–21, 122; Indian tribute, 20, 21–22, 26–27; during Mahuad presidency, 116, 118, 120, 122, 123, 124, 131, 135, 136, 163, 245; and oil, 35; value-added tax, 45, 118, 120–21, 124, 135, 163. *See also* Government spending
Technology, 47–48
Terán, Fausto, 208
Terán, Pablo, 206
Texaco Gulf consortium, 33, 36, 51, 56, 58, 60
Thailand, overvalued currency of, 150
Thatcher, Margaret, 44
Tigre, Daniel, 72
Torres, Lenín, 174
Torres, Sandino, 174, 175, 228
Torres Zapata, Wilson, 170
Tourism, 55, 56
Trans-Ecuadorean Pipe Line (SOTE), 33–34, 40, 44, 56, 57, 140–41, 216
Transportation, public, 38, 42, 86, 119
Tribunal of Constitutional Guarantees, 55
Tribute, Indian, 20, 21–22, 26–27
Tungurahua province, 3, 8, 9, 29–30, 67, 73, 138

Ulcuango, Ricardo, 72, 144, 177, 179
Ulloa, Juan and Antonio de, 19
Umajinga, César, 70
Unda, Hugo, 208, 225–26, 228
Union of Indian Campesinos of Cotacachi. *See* UNORCAC
Unions, labor, 88, 94, 119, 219–20, 231, 241; CTE, 61, 76; degree of unionization, 79; El Inca, 29, 61; FUT, 47, 73, 76, 93, 98–99; opposition to dollarization, 150, 152; relationship with CONAIE, 47, 61, 73, 76, 79, 221, 223, 245
United Fruit Company, 30
United Kingdom, denunciation of January 2000 coup, 190
United Nations, 5, 7, 187, 189–90; Human Development index, 159
United Pipeline Systems, 140
United Republican Party. *See* PUR
United Self-Defense Forces of Colombia. *See* AUC
United States: relations with Colombia, 144–46, 188, 229–30, 244; relations with Ecuador, 30, 33, 62, 63, 90, 96, 99, 100, 102–3, 107, 125–26, 140, 143, 144–46, 148, 153, 185, 187, 188–89, 207, 212, 214–15, 218, 229, 242, 243–44, 246
United Workers Front. *See* FUT
UNORCAC (Union of Indian Campesinos of Cotacachi), 70
Urbanization, 29, 37, 63
Urbina, José María, 27
Uruguay, trade bloc, 189
U.S. Agency for International Development, 56

Vallejo, César, 175
Vargas, Antonio, 69, 72, 135, 138, 143–44; on corruption, 211; during Levantamiento Indígena, 163, 164, 166, 168, 172–73, 176, 186, 187, 193–94, 197, 201–2, 235; after Levantamiento Indígena, 209, 212, 217, 233, 237, 245; on Mendoza, 193–94, 197, 201, 210; on Gustavo Noboa, 198, 210–11, 233–34; on Pachakutik, 76, 77; as pacifist, 177, 181, 195, 201, 203; relationship with Gutiérrez, 173, 176; relationship with Solórzano, 179; relations with Armed Forces, 168–69, 172–73, 174–75, 176–77, 193–94, 195–97, 201, 210, 225; on U.S. opposition to Junta of National Salvation, 189
Vega, Arcesio, 153
Velasco Garcés, Carlos, 86–87

Velasco Ibarra, José María, 30–31, 35–36, 108, 115; attitudes toward Indians, 28, 51–52, 71; on Ecuador as difficult to govern, 161; relations with Peru under, 126
Velásquez, Jacinto, 208, 235
Venezuela, 24–25, 41, 174, 180, 228, 237, 243
Verduga, Franklin, 109
Villacis, César, 175
Villacis, Luis, 70, 229
Villarreal, Rigoberto, 157
Villarroel, Jorge, 167, 170, 183, 195, 208, 224–25, 226
Viteri, Carlos, 69
Vivas, Juan José, 206, 221
War Academy, 175–76, 183, 184, 194
Wildlife Conservation International, 56
Women's Political Movement. *See* CPM
World Bank, 43, 45, 46, 87, 130, 140, 214, 215. *See also* IMF

Yandún, René, 209
Yasuni National Park, 55–56
Yépez, Ana Mariana, 133–34, 230
Yulee, Ramón, 132, 146

Zamora Chinchipe province, 4, 8, 10, 53, 73, 126, 179

Latin American Silhouettes
Studies in History and Culture

William H. Beezley and
Judith Ewell
Editors

Volumes Published

Brian Loveman and Thomas M. Davies, Jr., eds., *The Politics of Antipolitics: The Military in Latin America*, 3d ed., revised and updated (1996). Cloth ISBN 0-8420-2609-6 Paper ISBN 0-8420-2611-8

Dianne Walta Hart, *Undocumented in L.A.: An Immigrant's Story* (1997). Cloth ISBN 0-8420-2648-7 Paper ISBN 0-8420-2649-5

William H. Beezley and Judith Ewell, eds., *The Human Tradition in Modern Latin America* (1997). Cloth ISBN 0-8420-2612-6 Paper ISBN 0-8420-2613-4

Donald F. Stevens, ed., *Based on a True Story: Latin American History at the Movies* (1997). Cloth ISBN 0-8420-2582-0 Paper ISBN 0-8420-2781-5

Jaime E. Rodríguez O., ed., *The Origins of Mexican National Politics, 1808–1847* (1997). Paper ISBN 0-8420-2723-8

Che Guevara, *Guerrilla Warfare*, with revised and updated introduction and case studies by Brian Loveman and Thomas M. Davies, Jr., 3d ed. (1997). Cloth ISBN 0-8420-2677-0 Paper ISBN 0-8420-2678-9

Adrian A. Bantjes, *As If Jesus Walked on Earth: Cardenismo, Sonora, and the Mexican Revolution* (1998; rev. ed., 2000). Cloth ISBN 0-8420-2653-3 Paper ISBN 0-8420-2751-3

A. Kim Clark, *The Redemptive Work: Railway and Nation in Ecuador, 1895–1930* (1998). Cloth ISBN 0-8420-2674-6 Paper ISBN 0-8420-5013-2

Louis A. Pérez, Jr., ed., *Impressions of Cuba in the Nineteenth Century: The Travel Diary of Joseph J. Dimock* (1998). Cloth ISBN 0-8420-2657-6 Paper ISBN 0-8420-2658-4

June E. Hahner, ed., *Women through Women's Eyes: Latin American Women in Nineteenth-Century Travel Accounts* (1998). Cloth ISBN 0-8420-2633-9 Paper ISBN 0-8420-2634-7

James P. Brennan, ed., *Peronism and Argentina* (1998). ISBN 0-8420-2706-8

John Mason Hart, ed., *Border Crossings: Mexican and Mexican-American Workers* (1998). Cloth ISBN 0-8420-2716-5 Paper ISBN 0-8420-2717-3

Brian Loveman, *For* la Patria: *Politics and the Armed Forces in Latin America* (1999). Cloth ISBN 0-8420-2772-6 Paper ISBN 0-8420-2773-4

Guy P. C. Thomson, with David G. LaFrance, *Patriotism, Politics, and Popular Liberalism in Nineteenth-Century Mexico: Juan Francisco Lucas and the Puebla Sierra* (1999). ISBN 0-8420-2683-5

Robert Woodmansee Herr, in collaboration with Richard Herr, *An American Family in the Mexican Revolution* (1999). ISBN 0-8420-2724-6

Juan Pedro Viqueira Albán, trans. Sonya Lipsett-Rivera and Sergio Rivera Ayala, *Propriety and Permissiveness in Bourbon Mexico* (1999). Cloth ISBN 0-8420-2466-2 Paper ISBN 0-8420-2467-0

Stephen R. Niblo, *Mexico in the 1940s: Modernity, Politics, and Corruption* (1999). Cloth ISBN 0-8420-2794-7 Paper (2001) ISBN 0-8420-2795-5

David E. Lorey, *The U.S.-Mexican Border in the Twentieth Century* (1999). Cloth ISBN 0-8420-2755-6 Paper ISBN 0-8420-2756-4

Joanne Hershfield and David R. Maciel, eds., *Mexico's Cinema: A Century of Films and Filmmakers* (2000). Cloth ISBN 0-8420-2681-9 Paper ISBN 0-8420-2682-7

Peter V. N. Henderson, *In the Absence of Don Porfirio: Francisco León de la Barra and the Mexican Revolution* (2000). ISBN 0-8420-2774-2

Mark T. Gilderhus, *The Second Century: U.S.-Latin American Relations since 1889* (2000). Cloth ISBN 0-8420-2413-1 Paper ISBN 0-8420-2414-X

Catherine Moses, *Real Life in Castro's Cuba* (2000). Cloth ISBN 0-8420-2836-6 Paper ISBN 0-8420-2837-4

K. Lynn Stoner, ed./comp., with Luis Hipólito Serrano Pérez, *Cuban and Cuban-American Women: An Annotated Bibliography* (2000). ISBN 0-8420-2643-6

Thomas D. Schoonover, *The French in Central America: Culture and Commerce, 1820–1930* (2000). ISBN 0-8420-2792-0

Enrique C. Ochoa, *Feeding Mexico: The Political Uses of Food since 1910* (2000). Cloth ISBN 0-8420-2812-9 (2002) Paper ISBN 0-8420-2813-7

Thomas W. Walker and Ariel C. Armony, eds., *Repression, Resistance, and Democratic Transition in Central America* (2000). Cloth ISBN 0-8420-2766-1 Paper ISBN 0-8420-2768-8

William H. Beezley and David E. Lorey, eds., *¡Viva México! ¡Viva la Independencia! Celebrations of September 16* (2001). Cloth ISBN 0-8420-2914-1 Paper ISBN 0-8420-2915-X

Jeffrey M. Pilcher, *Cantinflas and the Chaos of Mexican Modernity* (2001). Cloth ISBN 0-8420-2769-6 Paper ISBN 0-8420-2771-8

Victor M. Uribe-Uran, ed., *State and Society in Spanish America during the Age of Revolution* (2001). Cloth ISBN 0-8420-2873-0 Paper ISBN 0-8420-2874-9

Andrew Grant Wood, *Revolution in the Street: Women, Workers, and Urban Protest in Veracruz, 1870–1927* (2001). Cloth ISBN 0-8420-2879-X (2002) Paper ISBN 0-8420-2880-3

Charles Bergquist, Ricardo Peñaranda, and Gonzalo Sánchez G., eds., *Violence in Colombia, 1990–2000: Waging War and Negotiating Peace* (2001). Cloth ISBN 0-8420-2869-2 Paper ISBN 0-8420-2870-6

William Schell, Jr., *Integral Outsiders: The American Colony in Mexico City, 1876–1911* (2001). ISBN 0-8420-2838-2

John Lynch, *Argentine Caudillo: Juan Manuel de Rosas* (2001). Cloth ISBN 0-8420-2897-8 Paper ISBN 0-8420-2898-6

Samuel Basch, M.D., ed. and trans. Fred D. Ullman, *Recollections of Mexico: The Last Ten Months of Maximilian's Empire* (2001). ISBN 0-8420-2962-1

David Sowell, *The Tale of Healer Miguel Perdomo Neira: Medicine, Ideologies, and Power in the Nineteenth-Century Andes* (2001). Cloth ISBN 0-8420-2826-9 Paper ISBN 0-8420-2827-7

June E. Hahner, ed., *A Parisian in Brazil: The Travel Account of a Frenchwoman in Nineteenth-Century Rio de Janeiro* (2001). Cloth ISBN 0-8420-2854-4 Paper ISBN 0-8420-2855-2

Richard A. Warren, *Vagrants and Citizens: Politics and the Masses in Mexico City from Colony to Republic* (2001). ISBN 0-8420-2964-8

Roderick J. Barman, *Princess Isabel of Brazil: Gender and Power in the Nineteenth Century* (2002). Cloth ISBN 0-8420-2845-5 Paper ISBN 0-8420-2846-3

Stuart F. Voss, *Latin America in the Middle Period, 1750–1929* (2002). Cloth ISBN 0-8420-5024-8 Paper ISBN 0-8420-5025-6

Lester D. Langley, *The Banana Wars: United States Intervention in the Caribbean, 1898–1934*, with new introduction (2002). Cloth ISBN 0-8420-5046-9 Paper ISBN 0-8420-5047-7

Mariano Ben Plotkin, *Mañana es San Perón: A Cultural History of Perón's Argentina* (2003). Cloth ISBN 0-8420-5028-0 Paper ISBN 0-8420-5029-9

Allen Gerlach, *Indians, Oil, and Politics: A Recent History of Ecuador* (2003). Cloth ISBN 0-8420-5107-4 Paper ISBN 0-8420-5108-2

Karen Racine, *Francisco de Miranda: A Transatlantic Life in the Age of Revolution* (2003). Cloth ISBN 0-8420-2909-5 Paper ISBN 0-8420-2910-9